THE M. & E. HANDBOOK SERIES

GEOLOGY

A. W. R. POTTER, B.Sc.

Senior Lecturer in Geology, The Polytechnic, Huddersfield

and

H. ROBINSON, B.A., M.Ed., Ph.D.

*Head of Department of Geography and Geology,
The Polytechnic, Huddersfield*

MACDONALD AND EVANS

MACDONALD AND EVANS LTD.
Estover, Plymouth PL6 7PZ

First published 1975
Reprinted in this format 1978

©

MACDONALD AND EVANS LIMITED

1975

ISBN: 0 7121 0719 3

Printed and bound in Great Britain by
Hazell Watson & Viney Ltd., Aylesbury, Bucks

PREFACE

THIS HANDBOOK on geology has been prepared to meet the need of students taking G.C.E. and other comparable examinations. It will be useful for students in Colleges of Technology undertaking introductory courses in geology. It should also be helpful to anyone who desires to have some basic knowledge of the physical environment of the Earth upon which we live.

The physical environment is the fundamental basis of our world and our very lives and all we do arise out of, and are made possible by, the very complex processes and interrelationships of land, air and water. Without any of these elements, there could be no Earth as we know it: the Earth would be as sterile, as lifeless and as motionless as the surface of the Moon. Geology is mainly concerned with the land surface and what lies below, but air and water and their influences also come into the study. It will, therefore, be of some interest to know something of the physical part of the Earth; for example, of the rocks which make up its surface, of the forces which lie below the surface, of the formation of mountains and plains, of how the water cycle works, and of the erosive and transportation activities of rivers, glaciers, the wind and the sea. There are chapters on fossils, which have provided very helpful clues for the elucidation of the ages of rocks and the sequence of geological events. There is a chapter on the applied aspects of geology, for geology is very much a practical science as well as being an academic study. Finally, a chapter has been incorporated on geological maps.

The book is meant to serve two main purposes. First, it seeks to provide a framework upon which the student can build and expand his geological knowledge. Experience shows that many students are unable "to see the wood for the trees," and these outlines are meant to help students, particularly those working mostly on their own and receiving little expert guidance, to a systematic understanding of the basic facts and principles of geology. Secondly, the book, by its orderly and succinct treatment, provides a useful summary for revision purposes. While the book attempts to cover the basic essentials, the text must be looked upon merely as a skeletal framework. Wider reading will give the student fuller and more detailed knowledge and a few books are recommended to the student for guidance; the following will be found useful:

H. H. Read and J. Watson: *Introduction to Geology*.

H. Robinson: *Morphology and Landscape*.

B. Simpson: *Rocks and Minerals*.

J. Kirkaldy: *Fossils in Colour*.

J. G. C. Anderson and T. R. Owen: *The Structure of the British Isles*.

M. J. Bradshaw and E. A. Jarman: *Reading Geological Maps*.

J. T. Greensmith: *Practical Geology for Schools*.

In addition to these the following series often provide valuable information in regional studies of Great Britain:

Regional Geology of Great Britain: a series of eighteen handbooks produced by the Institute of Geological Sciences.

Guides to the Geology Around University Towns, and Geology of Some Classic British Areas: produced by the Geologists Association.

British Landscape Through Maps: a series produced by the Geologists Association.

A guide to the answering of questions is given in Appendix II, and the student is advised to read this carefully and heed what it says. Appendix III gives some sample examination questions.

March 1975
A.W.R.P.
H.R.

NOTICE TO LECTURERS

Many lecturers are now using **HANDBOOKS** as working texts to save time otherwise wasted by students in protracted note-taking. The purpose of the series is to meet practical teaching requirements as far as possible, and lecturers are cordially invited to forward comments or criticisms to the Publishers for consideration.

P. W. D. REDMOND
General Editor

M & E HANDBOOKS

M & E Handbooks are recommended reading for examination syllabuses all over the world. Because each Handbook covers its subject clearly and concisely books in the series form a vital part of many college, university, school and home study courses.

Handbooks contain detailed information stripped of unnecessary padding, making each title a comprehensive self-tuition course. They are amplified with numerous self-testing questions in the form of Progress Tests at the end of each chapter, each text-referenced for easy checking. Every Handbook closes with an appendix which advises on examination technique. For all these reasons, Handbooks are ideal for pre-examination revision.

The handy pocket-book size and competitive price make Handbooks the perfect choice for anyone who wants to grasp the essentials of a subject quickly and easily.

CONTENTS

LIST OF ILLUSTRATIONS

LIST OF TABLES

CHAPTER I

THE SCIENCE OF GEOLOGY

INTRODUCTION

1. The meaning and scope of geology. The word geology comes from the Greek *ge* meaning the earth and *logos* a discourse; hence geology might be defined as the *science concerned with the description and understanding of the earth* from its creation to the present day. Geology, then, has the earth as its special field of investigation but deals more particularly with the crust. The aim of the geologist in the broadest sense is:

(*a*) to describe and interpret the surface physical features of the earth explaining at the same time their mode of origin; and

(*b*) to decipher and elucidate the history of the earth's evolution and of its past life from the records preserved in the rocks.

Geology deals largely with the study of minerals, rocks and fossils, not only as objects in themselves but as means to an end—the explanation of the historical development of the crust and its present physical features. The origin of rocks and their structural arrangement form a study which sheds much light upon the history of the earth, while the study of fossils provides much evidence about the history and evolution of life.

2. Geology and its relationship with other sciences. As a formal science geology is only some 200 years old but within the past couple of centuries tremendous advances have been made. The geologist calls upon the help of many other sciences to aid him in his work of detection and in providing answers to many of the problems which face him. Not only has he borrowed from other sciences but there is often, especially in his more specialised fields of study, considerable overlap with other sciences; for example, there is an aspect of geology called geophysics which combines geology and physics, while palaeontology, which studies fossil life, is clearly linked with biology. Figure 1 is a wheel graph which attempts to show the relationship of geology with the other sciences—cosmology, physiography, hydrology, meteorology, physics, chemistry, biology and economics. All these sciences make some contribution to geology.

Before we go on to note the different branches of geology which are recognised, it will be useful, and perhaps of interest, to trace briefly the origins and growth of geological science.

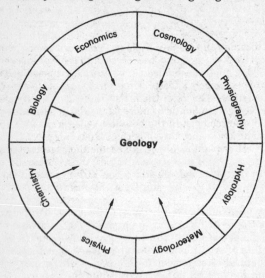

FIG. 1.—*The relationship of geology with the other sciences*. All make some contribution to geology and some have given rise to specialised aspects of geology, *e.g.* geophysics, geochemistry, geohydrology.

3. The growth of geological science. The beginnings of geological observations and deductions go back to the days of the early Greeks but the conclusions arrived at by classical writers were mere speculations. Nothing of real scientific value was arrived at until about two hundred years ago, although that Renaissance genius Leonardo da Vinci (1452–1519), who was an engineer and scientist as well as an artist, claimed that fossils were the remains of plants and animals and asserted that the rocks found at the tops of the Alpine Mountains were once under the sea.

During the period 1775–1825 the foundations of the science of geology were laid: many of the facts, principles and methods of modern geological science were assembled and laid down. The men chiefly responsible for laying down these foundations were the Scotsman Hutton, the Englishman Smith, the German Werner and the Frenchmen Lamarck and Cuvier. These five men may be said to be the pioneer founding-fathers of geology.

(a) *James Hutton* (1726–1797) published a work in two volumes—*Theory of the Earth*, in which he correctly interpreted the processes of erosion and deposition, the consolidation of sediments into beds of rock and the processes of folding and uplift. Hutton may be said to have established the principles of what has come to be known as physical geology and he is regarded as the founder of this branch of geology. Hutton believed that the changes in the earth's crust were largely the outcome of the action of fire and, as a consequence, his followers came to be known as *plutonists* (from Pluto, the god of the infernal regions).

(b) *William Smith* (1769–1839) was an English surveyor and civil engineer who made a study of sedimentary layers of rock and found that the various strata could be distinguished by the assemblages of fossils in them. He maintained that the age and sequence of rocks could be ascertained by their fossil contents. "Strata Smith", as he was dubbed, traced and recognised the various rock formations in Britain and in 1815 he drew and published the first *Geological Map of England and Wales with Part of Scotland*. Smith laid the foundations of stratigraphical geology, another important subdivision of the science.

(c) *Abraham Gottlieb Werner* (1750–1817) was a professor in the Academy of Mining at Freiburg in Germany. He was a great teacher who fired his students with an enthusiasm for geology and so, in this way, exerted a powerful influence upon the early development of geology. Werner, however, believed in the aqueous origin of igneous rocks; this was an old-fashioned but erroneous belief. His followers were known as *neptunists* (after Neptune, the god of the sea). There was a prolonged and violent controversy between the Plutonists and the Neptunists as to the origin of the class of rocks called igneous. Werner was the founder of economic geology.

(d) *Jean Baptiste de Lamarck* (1744–1829) was a distinguished French biologist who investigated fossil remains and used them to work out the history of the earth and the succession of events. He was one of the first to recognise the great age of the earth. More important, he was one of the originators of the theory of evolution and opposed the then held ideas of catastrophic destruction and repeated creations of life.

(e) *Georges Cuvier* (1769–1832) was a Professor of Comparative Anatomy in Paris and described fossil mammals which were dug up in France. He was a founder of palaeontology, the study of fossils. He rejected Lamarck's correct and important conclusions about the continuity of life on the earth; nevertheless he made important contributions to the early development of geology.

DIVISIONS OF THE SCIENCE

4. The branches of geology. The scope of geology is large: geology literally means the science of the earth and as such it seeks to provide an understanding of earth processes and earth materials. In addition the geologist is concerned with the history of the earth as revealed by those processes and materials. Thus it is often convenient to sub-divide the subject into a number of areas of study all of which are interdependent (Fig. 2).

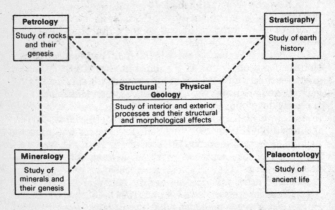

| Petrology |
| Study of rocks and their genesis |

| Stratigraphy |
| Study of earth history |

| Structural   Physical Geology |
| Study of interior and exterior processes and their structural and morphological effects |

| Mineralogy |
| Study of minerals and their genesis |

| Palaeontology |
| Study of ancient life |

Fig. 2.—*The principal elements of geology.*

The main branches are:

(a) Mineralogy and petrology: the study of minerals and rocks.

(b) Palaeontology: the study of ancient life.

(c) Physical geology: the study of earth processes and their resultant effects.

(d) Stratigraphy or historical geology: the study of earth history.

(e) Applied geology: a term used to group the varied geological investigations associated with human economic activity.

Some would add other divisions. For example during the last three decades in particular, great advances in technology have led to a rapid increase in the importance of the related fields of *geochemistry* and *geophysics*. Many techniques which have contributed greatly to geological knowledge in recent years have come from these particular areas of study.

5. Mineralogy and petrology. These study the materials of the earth's crust, *i.e.* the rocks and the minerals which make up the rocks.

(a) *Mineralogy* studies the minerals which form the rocks. A specialised aspect of mineralogy is crystallography, which studies crystals.

(b) *Petrology* studies the origin, occurrence and classification of rocks.

6. Palaeontology. Palaeontologists investigate all aspects of fossilised life found in rocks. This has produced a knowledge of the evolutionary changes of life throughout earth history and has enabled us to interpret the environmental conditions of the past.

7. Physical geology. This branch deals with the processes which operate within and on the surface of the earth and which continually modify its form. Perhaps two sub-divisions may be recognised:

(a) *Structural geology*—the study of interior and crustal structures, and the internal processes which produce them.

(b) *Geomorphology*—deals with the shapes of landforms and the processes which act upon them.

8. Stratigraphy. This fourth section of geology studies the procession of changes throughout the history of the earth. It includes such elements as:

(a) *Geochronology*—concerned with the dating of rocks and the erection of an accurate geological time scale.

(b) *Palaeogeography*—studies the geographical conditions obtaining during former geological ages and attempts to reconstruct past geographies.

9. Applied geology. The practical applications of geological knowledge are of vital importance to man in such areas of study as:

(a) *Mining geology* or the study of mineral deposits and of the means by which they can be found.

(b) *Petroleum geology* which deals with the problems of locating and extracting petroleum and natural gas.

(c) *Hydrogeology* or the study of the occurrence, extraction and use of underground water supplies.

(d) *Engineering geology* which deals with the relations of geology to engineering operations. Of particular importance in this work are the studies of soil and rock mechanics which supply data on the behaviour of rock foundations under the stress of constructional works.

(e) *Environmental geology.* Geology is able to play an important part in the evaluation of man's interactions with his environment. The mapping, prediction and control of environmental hazards such as earthquakes, volcanic eruptions and landslides provides just one example of the many researches occurring in this sphere at present.

10. The importance of geology. The study of geology is important for numerous reasons; here are a few of the more obvious ones:

(a) It helps us to understand the origin, nature and structure of the earth as a planet and especially of the continents and oceans of the earth's crust.

(b) It helps to explain the varied character of the landscape and how the many different landforms have come into being.

(c) It is important in the study of soils, for the rocks form the raw material, so to speak, of soil and are partly responsible for the development and for the physical and chemical characteristics of soil.

(d) It is important in connection with water resources because the rocks, through their character and structure influence run-off and the percolation of water underground; rocks may also affect the quality as well as the quantity of water made available to man.

(e) Rocks provide us with our mineral wealth and the expert geologist is often able to locate precise areas where minerals are likely to be found. Prospecting is very closely bound up with a knowledge of rocks and rock structures.

(f) The availability of building materials—stone, brick, cement—is dependent upon the geology of an area to a very considerable extent; in the past, regional styles of architecture were very closely tied to the constructional materials available.

(g) A knowledge of geology can help the engineer, enabling him to anticipate and avoid the danger and calamities which sometimes overtook his forerunners who did not have expert geological knowledge to help them.

(h) A knowledge of geology helps us to be aware of and to forecast, though not with any precise accuracy, certain natural calamities such as earthquakes and volcanic activity.

(i) The study of geology, more particularly through its fossil record, helps us to understand the process of natural evolution.

PROGRESS TEST 1

1. Describe the special field of investigation of the science of geology. (1)

2. Name five of the founders of geology and indicate the contributions made by any two of them to geological science. (3)

3. What are the chief branches of geology? Describe briefly the fields or areas of investigation of each of these branches. (4–9)

4. With what does the branch of applied geology concern itself? (9)

5. Give several reasons why the study of geology is important. (10)

6. Briefly explain the following terms: Plutonists, palaeontology, geomorphology, hydrogeology, geochronology. (3, 6, 7, 8, 9.)

CHAPTER II

THE GEOLOGICAL TIME-SCALE

THE PRINCIPLES OF STRATIGRAPHY

Whilst the petrologist usually adopts a classification of rocks based upon their origin and composition, an alternative classification may be made according to their age. The branch of geology known as *stratigraphy, which studies the succession of rocks*, needs to know the age of rocks and to use terms to describe their age. The geologist, accordingly, has devised a chronological table which enables him to show the sequence in which the different strata found in the crust were formed. In building up this chronological table the geologist has made use of a series of basic geological principles. In broad outline they involve the following:

1. The principle of uniformitarianism. This is often explained as meaning "the present is the key to the past." It was a method largely initiated by James Hutton in the eighteenth century. It assumes that geological processes and their products forming on the earth today are essentially the same as those of the past. Thus we can interpret the rocks that were formed in the past by drawing an analogy with their present day counterparts. In fact conditions and processes were not exactly the same in past times as today, but nevertheless this concept has been a most useful aid, particularly in determining palaeoenvironments (*i.e.* environments in the geological past) from faunal and sedimentological evidence.

2. The relative dating and correlation of rocks. Dating of rocks is basically of two kinds—*relative dating* and *absolute dating*. All of the early pioneer work carried out in geology involved the relative dating of strata—that is, the exact time of their formation in terms of years was not known. Utilised for this were the following:

(*a*) *The law of superposition*. This simply states that in a stratified sequence each bed is younger than the one it lies upon and older than the one which lies upon it. This simple fact allowed the establishment of a geological sequence of beds whose ages relative to each other were known. Of course the law cannot apply where tectonic disturbances have inverted the strata. The direction of *younging*, *i.e.* the direction in which the beds are getting younger, of such disturbed sequences can sometimes be determined from sedimentary structures or from cleavage-bedding intersections.

(b) *Methods of correlation*. Having established the relative ages of a series of strata, it was necessary to have a way of recognising them when they occurred in widely separated outcrops (*i.e.* of correlating them). One possible solution was to use their *lithological* characteristics (*e.g.* whether the bed is a sandstone or a shale of a particular thickness and colour and whether it has particular sedimentary structures within it). The drawback to this kind of correlation is that lithologies are often found to change laterally. Thus, for instance, a sandstone often changes into a siltstone and then into a shale over a very short lateral distance. In spite of such difficulties lithological mapping remains one of the basic methods of surveying today.

It was William Smith and other workers in the early nineteenth century who found that fossils provided the key to the solution. Once it was clear that any species lives on earth for only a limited period and never reappears, it followed that its fossilised remains would only be found in a limited thickness of strata. Thus a particular stratum or group of strata could often be identified by the fossil assemblage it contained. Ultimately this concept led to the establishment of *faunal zones*. A zone may be defined as strata deposited during a period of time (a secule) during which a particular faunal assemblage existed. It is usually named from one of the characteristic species of the assemblage which is then called the *zone fossil*. To be ideal for stratigraphical purposes a species should fulfil the following conditions:

(i) It should have a wide geographical distribution which is rapidly attained. Free-swimming forms tolerating a variety of conditions are most suitable for this.

(ii) It should have a short vertical range—*i.e.* they should live

TABLE I: STRATIGRAPHICAL CHART

Era	Period	Epoch	Age of beginning (Millions of years)	Fauna, Flora	Major structural episodes
Cainozoic	Quaternary	Recent	c. 0·01	The development and dominance of *Homo sapiens* takes place. Great climatic fluctuations are reflected in vegetational fluctuations as a series of glaciations affects many Northern Hemisphere countries.	
		Pleistocene	c. 2		
	Tertiary	Pliocene	c. 7	Mammals become dominant among vertebrates. Primate evolution leads to the first primitive men at the close of this period. Vegetation takes on modern appearance. Forests, very extensive initially, are partially replaced by grasslands towards the end. Lamellibranchs and gastropods proliferate.	Main Alpine Episode
		Miocene	26		
		Oligocene	38		
		Eocene	54		
		Palaeocene	64		Laramide Phase

				—general world-wide elevation
Mesozoic	Cretaceous	136	Large reptiles dominate the vertebrates although the early mammalian and bird evolution occurs here. Conifers and ferns common and flowering plants in Cretaceous. Ammonites, brachiopods and echinoderms important among invertebrates.	*Main Hercynian Episode*
	Jurassic	190—195		
	Triassic	225		
Palaeozoic	Permian	280	Fish become abundant and amphibians evolve from them at end of Devonian. Reptiles develop during Carboniferous as spore-bearing plants colonise the land. Molluscs, brachiopods and corals are common invertebrates.	
	Carboniferous	345		
	Devonian	410±10		*Main Caledonian Episode*
	Silurian	440±10	Trilobites and graptolites most important invertebrates although all the phyla present. Jawless fish appear and increase towards the end.	
	Ordovician	530±10		
	Cambrian	570±10		
Azoic or Pre-Cambrian eras		Oldest rocks 3787±85	Life poorly preserved. It is thought that primitive plant life of an algal type was followed by the development of soft-bodied animals (e.g. worms, etc.).	*Several Mountain Building Episodes* 1100—850 2000—1600 3300—2500

only for a very limited period and thus appear in a very limited thickness of strata.

(*iii*) It should be numerous.

(*iv*) It should be suitable for good preservation.

3. Absolute (radiometric) dating. Although various methods were adopted in an effort to determine the absolute ages of strata, it was not until the advent of radiometric dating during this century that much progress was made. The method is based on the fact that naturally occurring radio-active isotopes of some elements decay slowly, and at a measurable steady rate. By determining how much decay has taken place in a specimen, it is possible to deduce when the breakdown started and hence the age of the specimen. In fact there are many practical difficulties associated with this, and no single determination carried out in isolation can be regarded as reliable. Yet, over the years, large numbers of determinations have enabled us to establish the figures which are shown in the stratigraphical chart (Table I).

THE DIVISION OF GEOLOGICAL TIME

4. Forms of animal life as a basis. Pre-Cambrian time extended well over 3,000 million years and our knowledge of events in these early times is only slight. Much of the evidence of what happened has been obscured by later events and this leads to great difficulty in interpretation.

More recent geological time—about the last 600 million years, which accounts for only one-seventh of total geological time— is divided into three eras and these time divisions are based upon *the forms of animal life which are represented by the fossil remains in the rocks*. In the "era names," the ending "zoic," which is derived from the Greek word *zoon* meaning "animal," may be loosely interpreted as "the form of life."

The three eras are:

(*a*) *Palaeozoic*—"ancient" (*palaios*) forms of life;
(*b*) *Mesozoic*—"intermediate" (*mesos*) forms of life;
(*c*) *Cainozoic*—"recent" or "new" (*kainos*) forms of life.

Alternative names for these eras are: Primary, Secondary, Tertiary. However, we should note that it is now common practice among geologists to divide the Cainozoic Era into two parts: the Tertiary and the Quaternary. The Quaternary covers the past two million or so years during which time man has evolved.

5. Major divisions of geological time. The geologist, then, divides the vast span of geological time into four major divisions but these, as in the case of the historian's divisions of convenience, are not of equal duration.

(a) *The Pre-Cambrian Eras.* These cover a vast length of time, something of the order of 4,000 million years, and during Pre-Cambrian times many series of rocks must have been laid down. Some of these very old sedimentary rocks are still preserved in the extreme north-west of the British Isles, *e.g.* the Torridonian sandstones. As we have already indicated, very little fossil evidence of life-forms exists although simple plants and animals both made their appearance during quite early Pre-Cambrian times.

(b) *The Primary or Palaeozoic Era.* This era extended over a period of some 345 million years so far as we can tell. The rocks which were formed comprised sediments of all kinds while in some areas there were phases of igneous activity. The sedimentary strata contain fossils of "ancient" forms of life such as spineless creatures, early fishes and algae, while towards the end of the era the first land plants appeared. The Palaeozoic era ended approximately 225 million years ago.

(c) *The Secondary or Mesozoic Era.* This era extended over a period of approximately 160 million years and drew to a close about 64 million years ago. It is a period of "intermediate" forms of life when reptilian forms were especially numerous and plant life was being elaborated. Many groups of plants of modern type, such as flowering plants, begin to emerge towards the end of the era.

(d) *The Cainozoic Era.* Most of this era is occupied by the Tertiary for the Quaternary covers only the last two million years or so. During the Cainozoic Era "recent" forms of life emerged and this is the age essentially of flowering plants and mammals. This era has been marked by pronounced mountain-building activity and the earth as we know it today was marked out during this time.

6. The subdivisions of the eras.. The above eras are basically distinguished by the particular forms of life that were in existence during the earth's past. The fossil record indicates a gradual evolution from the most simple and primitive forms—when, in fact, it is often difficult to tell whether a life form is an animal or a plant—to the most complex and highly developed forms.

During each era, however, many kinds of rock were formed and tremendous thicknesses of sediments were laid down. Each of these contains fossils characteristic of their time. It becomes necessary, therefore, to have a more detailed time division. The geologist divides the eras into smaller time units and such subdivisions in decreasing order of magnitude are termed *periods* and *epochs*. Groups of rocks deposited within periods form *systems*: those within epochs *series*. The smallest unit of strata is termed a zone.

This breakdown is illustrated in Table II.

TABLE II: SUB-DIVISION OF THE MESOZOIC ERA

Era	Period (and system)	Lithological formation	Estimated Representative thickness	
			feet	*metres*
MESOZOIC	Cretaceous	Chalk	1,000	305
		Upper Greensand	100	30
		Gault Clay	300	90
		Lower Greensand	500	150
		Wealden Clay	800	240
		Hastings Beds	650	200
	Jurassic	Purbeck Beds	250	75
		Portland Beds	200	60
		Kimmeridge Clay	800	240
		Corallian Beds	50	15
		Oxford Clay and Kellaways Beds	400	120
		Cornbrash	20	6
		Great Oolite Series	140	40
		Inferior Oolite Series	150	45
		Lias	800	240
	Triassic	Rhaetic	40	12
		Keuper	1,800	545
		Bunter	1,000	305

It should finally be emphasized, however, that not all of the systems shown are represented in any one locality. There are often breaks in the stratigraphical record which are marked by *unconformities*. These may be angular discordances where one stratum *oversteps* a number of tilted or folded strata beneath it (*see* Fig. 3). *Overlap* is commonly found associated with such angular unconformities. It occurs where a deepening sea has

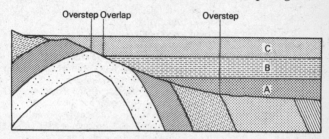

Fig. 3.—*Section illustrating overstep and overlap* (*e.g.* bed C overlaps bed B and oversteps the folded strata).

spread sediments over a wider and wider area. Thus each stratum laid down occupied a slightly greater area than the one before and is said to overlap it. Beds laid down in a shrinking sea exhibit the reverse effect, *i.e.* are deposited over a gradually reducing area: this is termed *offlap*. Small breaks in sedimentation are often not marked at all by angular discordance, and can only be recognised from fossil or sedimentological evidence. They are called *non-sequences*.

7. Nomenclature. The various systems are named in different ways. The terminology used is very mixed, but the names illustrate the gradual growth of geology as a science in both time and space. The names of the periods or systems are derived in five main ways:

(*a*) By describing the characteristic rocks, *e.g.* Cretaceous—chalk, Carboniferous—coal bearing.

(*b*) By describing the nature of the fossil contents, *e.g.* Oligocene—a few fossils in recent formation, Holocene—entirely recent.

(*c*) By indicating the number of divisions within the system, *e.g.* Triassic implying a threefold division.

(*d*) By using the place-names of the areas where the rocks were first recognised and studied, *e.g.* Devonian from Devon, Permian from the town of Perm in the Soviet Union.

(*e*) By using ancient British tribal names, *e.g.* Silurian after the Welsh tribe of the Silures, Ordovician after the Ordovices.

PROGRESS TEST 2

1. Briefly describe the basic principles used by the geologist to help him build up his chronological table. (1–3)

2. When the relative ages of a series of rock strata become known, it is necessary for the geologist to have a way of recognising them when they occur in widely separated outcrops: how does he do this? (2)

3. Briefly explain the following: the principle of uniformitarianism, the law of superposition, faunal zones, radiometric dating. (1–3)

4. What are the major divisions of geological time? How are their names derived? How long approximately did each last in years? (4, 5)

5. How does the geologist sub-divide the great eras of geological time and what different kinds of sub-divisions does he recognise? (6)

6. Write a brief account of the derivation of the nomenclature used to describe the various geological systems. (7)

7. Briefly explain the meaning of the following terms: younging, secule, overlap, non-sequences. (2, 6)

CHAPTER III

MINERALS AND THEIR PROPERTIES

THE NATURE OF MINERALS

1. Elements. All matter is made up of a number of substances and these substances, the simplest the chemist can obtain, are called elements. *Elements are substances in their most elementary form which cannot be split up into simpler substances by chemical means.* The number of elements known at the present day number just over 100. The elements may be roughly divided into two classes, metals and non-metals, but there is no hard-and-fast line between them and there are some elements known as metalloids, *e.g.* arsenic, which combine characteristics of both classes.

Of the elements that are known a few are abundant but many are extremely rare and of little importance to the mineralogist. It has been estimated that the crust of the earth is composed, preponderantly, of the following eight elements:

oxygen	46·71 per cent	calcium	3·65 per cent
silicon	27·69 per cent	sodium	2·75 per cent
aluminium	8·07 per cent	potassium	2·58 per cent
iron	5·05 per cent	magnesium	2·08 per cent

Over 98 per cent of the earth's crust is made up of these eight elements. None of these, however, with the exception of iron, exists on its own. Some of the other elements may exist in nature, making minerals by themselves, as do, for instance, gold, platinum, copper, sulphur, and carbon (in the form of graphite or diamonds). In most cases, however, two or more elements are chemically combined and when such a combination takes place then the elements lose their identities as such.

2. Definition of minerals. The materials constituting the earth's crust are, with a few exceptions, minerals. It is true that the term mineral is often loosely used to describe any substance of economic value that is extracted from the earth, *e.g.* coal, iron ore, oil, gravel, brick-earth. From a scientific point of view, however, this is inadequate and unacceptable: the geologist needs to be more precise, more scientific. To the geologist the term mineral is restricted to:

(a) substances of inorganic nature;
(b) substances having a definite chemical composition;
(c) substances possessing definite physical properties.

A mineral may be defined, therefore, as *a homogeneous substance of definite chemical composition or one that varies within small limits, and of constant properties found ready-made in nature and not a product of life.* According to such a definition, a mineral must be a naturally occurring inorganic substance of the same nature throughout and its composition must be so definite that it can be expressed by a chemical formula, *i.e.* it has a definite atomic structure.

It will be clear from this that:

(a) many substances of organic origin, such as coal, mineral oil, guano, are not minerals, partly because they are organic and partly because they are of variable composition;
(b) artificial substances such as laboratory and furnace (*i.e.* man-made) products, *e.g.* glass, brick, cement, etc., are not minerals since they are not natural.

Virtually all minerals are solid substances: only water ordinarily exists in liquid form. Altogether, about 1,500 mineral species are known; to these, and their varieties, many thousands of names have been given.

3. Mineral compounds. While, as we have said, a few elements may occur in their elemental state forming minerals, *the greater number of minerals found in rocks occur as compounds.* It should be understood that a chemical compound is something more than a mere mixture. Compounds may be described as pure substances, made up of two or more elements, which are formed as a result of chemical change.

Compounds differ from mixtures in several important ways:

(a) The elements which constitute a compound are combined in strictly definite proportions by weight whereas mixtures are not.
(b) The components of a mixture can be separated mechanically, but a compound cannot be so easily split up.
(c) The properties of a compound may be very different

from those of its constituent elements, whereas a mixture normally has the properties of its components.

(*d*) When a compound is formed heat is either absorbed or released but this does not normally occur when substances are merely mixed.

4. Examples of combinations. The simplest compounds consist of two elements combined together. For example, oxygen and silicon, a gas and a solid respectively, may be united to form the hard, common mineral quartz. Compounds of oxygen with another element are called oxides and form a very important class of minerals. The sulphides are compounds of elements with sulphur, *e.g.* galena (lead sulphide). The chlorides are compounds of chlorine with another element, *e.g.* halite (sodium chloride or common salt).

Two or more elements may combine. For example, the three elements of carbon, calcium and oxygen may unite to form the mineral calcite: four elements, potassium, aluminium, silicon and oxygen may combine chemically to form the mineral orthoclase (a felspar); and some minerals are even more complicated in their composition.

The oxides of metals are, in general, basic and frequently combine with water to form bases, *e.g.* caustic soda ($NaOH$). The oxides of non-metals are acidic and most of them dissolve in water to form acids, *e.g.* sulphuric acid (H_2SO_4). Basic and acid oxides frequently combine together forming compounds which contain three or more elements. The metals iron, calcium, magnesium, aluminium and potassium form the basic oxides known respectively as the oxide of iron, lime, magnesia, alumina and potash. On the other hand, the non-metals, silicon, carbon and sulphur form acid oxides. The compounds of these two sets of oxides with one another are known as silicates, carbonates, sulphates, etc., *e.g.* silicate of iron, carbonate of magnesia and sulphate of lime.

5. Minerals and rocks. *Minerals are compounds of their constituent elements; rocks are mixtures of their component minerals.* Thus the mineral quartz is a compound of the elements silicon and oxygen, whereas the rock granite is a mixture of several minerals—quartz, felspar, mica, etc. An examination of rock samples shows that in most cases rock consists of a mixture of various minerals: indeed, the heterogeneous rock may be taken to pieces and the minerals that compose it separated out.

The majority of rocks are made up of aggregates of minerals, though some, the organic rocks, such as coal, peat and guano, are not. Certain rock formations are made up essentially of but one mineral in the form of numerous individual grains which have become compacted or cemented together, *e.g.* pure sandstone which may contain only grains of quartz (oxide of silicon) or pure limestone which may consist wholly of calcite (carbonate of lime).

Most of the more commonly known rocks, however, are made up of two or more minerals which are bound together mechanically. Inspection of a piece of granite, for instance, shows that it contains several distinct mineral species which are easily distinguishable: hard, clear, glassy grains of quartz; hard, whitish or pinkish grains, often with smooth faces, of felspar; and small silvery white or black, soft, scaly flakes of mica.

6. The common rock-forming minerals. Most of the rock-forming minerals consist of oxides, chlorides, and sulphides, or of silicates, carbonates and sulphates. The most common of the rock-forming minerals belong to the last class. The most important rock-forming minerals are:

(*a*) oxides of silicon (quartz) and iron (magnetite and haematite);

(*b*) sulphide of iron (pyrites);

(*c*) chloride of sodium (halite);

(*d*) silicates

  (*i*) anhydrous, such as felspar, olivine,

  (*ii*) hydrous, such as chlorite, serpentine;

(*e*) carbonate of lime (calcite and aragonite), with magnesia (dolomite);

(*f*) sulphate of lime (gypsum).

PROPERTIES OF MINERALS

7. General characteristics. As we have just indicated all minerals have a distinctive chemical composition but they possess, in addition, certain physical properties which are peculiar to themselves. These physical properties are classified in the following groups:

(a) *Characteristics depending upon light* such as colour, lustre, transparency, translucency, phosphorescence and fluorescence.

(b) *Characteristics depending upon certain senses*, such as taste, odour and touch.

(c) *Characteristics depending upon the state of aggregation*, *i.e.* the atomic structure of the mineral, such as form, hardness, tenacity, fracture, cleavage, etc.

(d) The *specific gravity* of minerals.

(e) *Characteristics depending upon heat*, such as fusibility.

(f) *Characteristics depending upon magnetism, electricity and radioactivity*.

The more important of these properties are: crystalline form, cleavage, specific gravity, hardness, lustre, colour and streak. For present purposes these properties are sufficient for the identification of the more common minerals.

In a broad fashion we can divide these properties into two groups: optical properties, or those which can be distinguished simply by looking at the mineral (*see* 8–11), and the other, or physical, properties (*see* 12–17).

8. Colour. Minerals show a great variety of colour and often the colour of a mineral is its most striking property. But colour is very variable and so cannot be relied upon with any degree of certainty. It is unfortunate for purposes of identification that many minerals vary widely in colour: even within the same species specimens are to be found having different colours, *e.g.* quartz, which is commonly colourless or white, may be rose, violet, green, yellow or brown. Some minerals, however, possess distinctive colours as inherent characteristics which seldom fail and can almost always be relied upon, *e.g.* azurite (blue), malachite (green), cinnabar (brick-red), pyrite (brass-yellow). Many minerals in their pure form are colourless or white, e.g. quartz, gypsum, calcite, halite.

Some colours are due to the physical state of the mineral rather than its chemical content; for example:

(a) *opalescence*, due to thin layers of air in the mineral;

(b) *iridescence*, due to interference of light rays;

(c) *schiller*, a nearly metallic lustre, due to thin films of other mineral matter which break up light;

    (d) *tarnish*, resulting very often from exposure to air which causes oxidation.

9. Streak. The streak of a mineral is the colour of its powder. Certain minerals exhibit a different colour in powdered form from the mineral in mass. A sample of powdered mineral may be easily obtained by rubbing a specimen on a piece of unglazed porcelain (so-called "streak-plate") or, alternatively, by scratching a mineral with a file. The streak so obtained may be characteristic of a mineral and this greatly aids the identification of a species, *e.g.* black hematite gives a red streak, black limonite a yellowish-brown streak, copper pyrites a brown streak, cinnabar a scarlet streak, and sphalerite (zinc blende) a yellow streak.

10. Lustre. Lustre is the appearance of the surface of a mineral independent of its colour. *Lustre is the reflection given by the surface*; it varies considerably depending upon the amount and type of reflection of light. Some minerals reflect a great deal of light, others very little or none. Lustre is often more or less characteristic of a mineral. Six main kinds of lustre are recognised :

    (a) *Metallic*—when a mineral looks metallic and is silvery or brassy such as pyrites, gold, galena, graphite; when feebly displayed it is termed sub-metallic, *e.g.* chromite, cuprite.

    (b) *Vitreous*—when a mineral is glassy, like broken glass, *e.g.* quartz, halite; when it is less well-developed the term subvitreous is applied, *e.g.* in the case of calcite.

    (c) *Resinous*—when a mineral has the lustre of resin; amber, opal and some kinds of sphalerite have a resinous lustre.

    (d) *Pearly*—when a mineral possesses the lustre of a pearl; talc, mica, brucite and selenite show a pearly lustre.

    (e) *Silky*—when a mineral looks like silk; this kind of lustre is peculiar to minerals having a fibrous structure, *e.g.* satin-spar, the fibrous form of gypsum, and the variety of asbestos, amianthus.

    (f) *Adamantine*—a mineral possessing the clear, brilliant lustre of a diamond.

Minerals with no lustre are described as dull.

11. Diaphaneity. This property relates to the transmission of light through a thin piece of a mineral. Three degrees of diaphaneity are distinguished:

(a) *Transparent*—a mineral is said to be transparent when an object can be seen clearly through it.

(b) *Translucent*—if some light passes through it, but objects are not distinguishable, the mineral is translucent.

(c) *Opaque*—if no light passes through the mineral then it is said to be opaque.

All the above characteristics can be readily detected by the eye and, therefore, they are termed optical properties. It should be understood, however, that optical properties are also physical properties. The more common physical properties used to distinguish minerals are specific gravity, hardness, crystalline form, cleavage and fracture.

12. Specific gravity. The specific gravity of minerals, that is their *weight in proportion to that of an equal volume of water*, is a valuable means of detecting them. Though mainly a laboratory test, a simplified determination can be carried out in the field by means of a balance (so-called Walker's Balance after its maker), but a rough estimate can be obtained by feeling the weight of a mineral sample in the hand.

The range of specific gravity runs from less than 1 to about 23, the average of all minerals being about 2·6. It is important to note the specific gravity of a mineral specimen undergoing examination because it often helps the recognition of the species; for example, galena can be recognised immediately by its high specific gravity which is 7·5. Silver, also, by its extra heavy weight, 10·5, is readily recognised. Platinum, when pure, has a specific gravity of over 21, making it one of the heaviest known substances.

13. Hardness. An important criterion for the recognition of minerals is hardness: in this case hardness refers to the degree of resistance which a smooth mineral surface offers to abrasion or scratching. Scarcely any two minerals are exactly alike in hardness, but for practical purposes a generally adopted scale recognises ten degrees of hardness: this is known as *Mohs' scale*.

1. Soft, greasy feel and easily scratched by the finger nail, *e.g.* talc.

2. Just capable of being scratched by the finger nail, *e.g.* gypsum.

3. Capable of being scratched by a pin or copper coin, *e.g.* calcite.

4. Not scratched by a pin but scratched by a steel knife blade; the mineral will not scratch glass, *e.g.* fluorite.

5. The mineral just scratches common glass and is scratched by a knife, *e.g.* apatite.

6. Barely marked by a well-tempered steel knife: the mineral scratches common glass easily, *e.g.* orthoclase felspar.

7, 8, 9 and 10. Harder than most ordinary common substances and represented in order of hardness by quartz, topaz, corundum and diamond.

14. Crystalline form. All minerals, if free to form without interference and under suitable conditions, will assume a crystalline shape. Some minerals show their crystalline form quite clearly but others have such microscopic crystals that to the naked eye they do not appear to be crystalline. *Each mineral has only one characteristic crystal form*, but a crystal of any mineral may assume various shapes as a result of more rapid growth in one direction than another. Thus quartz always crystallises in hexagonal prisms terminated by hexagonal pyramids but the prisms may be thick and stumpy or long and thin according to the conditions under which the crystals grow.

As the crystal form depends ultimately on the arrangement of the atoms in the mineral molecule, some of its characterising features, such as the angles between the faces of the crystal, the direction of cleavages, *i.e.* the way it splits, and the optical properties of the mineral, are absolutely constant and capable of the finest degree of precision in their measurement.

All crystals can be classified under one or other of seven groups or systems:

(a) *cubic*
(b) *tetragonal*
(c) *hexagonal*
(d) *trigonal*
(e) *orthorhombic*
(f) *monoclinic*
(g) *triclinic.*

Figure 4 shows the different crystal forms.

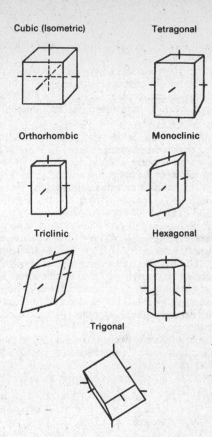

FIG. 4.—*The crystal systems.* 1. Cubic (Isometric): all axes are equal and are at right angles to one another. 2. Tetragonal: the three axes are at right angles but one is longer (or shorter) than the other pair. 3. Orthorhombic: the three axes are at right angles, but unequal: all the faces are oblong. 4. Monoclinic: the three axes are all unequal and one of them is not at right angles to one of the others. 5. Triclinic: none of the three axes is equal or at right angles. 6. Hexagonal: four axes, three of which are horizontal making angles of 60° with each other and are in the same plane and one of which is vertical and longer (or shorter) than the other three but is at right angles to the plane containing them. 7. Trigonal: some authorities do not separate this system from the Hexagonal for the same axes are used.

15. Cleavage. Many crystals and crystalline substances have a marked tendency to split easily in certain well defined directions giving more or less smooth faces; this is known as cleavage and is an important property of minerals. The splitting or cleavage takes place along planes of weakness, *i.e.* along planes of weaker molecular cohesion which lie parallel to the faces of the crystal.

The degree of cleavage varies and may be classed as being perfect, good or poor. An extreme form of cleavage is seen in the case of mica which is perfect: so strong is the cleavage in one direction that mica can be split into a number of thin plates as thin as paper. In similar tabular minerals perfect or almost perfect cleavage is common. Quartz, on the other hand, provides a good example of very poor cleavage.

There may be more than one direction in which a mineral splits; these are termed cleavage directions. If three are present regular geometrical fragments will be produced when the mineral is broken. The number of cleavage directions exhibited by common minerals is illustrated by the following: mica, one; felspar and amphibole, two; calcite and galena, three; fluorite, four; sphalerite, six. Sometimes the degree of cleavage varies with direction: for example, felspar has two cleavages, one good and a rather poorer one.

In the study of minerals careful attention should be given to cleavage whenever it occurs, since certain minerals always show particular cleavage directions.

16. Fracture. Different from cleavage is fracture. By fracture is meant the nature or character of the surface of a mineral when it is freshly broken. The manner of this fracture may be characteristic of particular minerals. The following terms are used to describe the type of fracture:

    (a) *conchoidal:* shell-like, curved concave or convex fractures, *e.g.* flint;
    (b) *sub-conchoidal:* e.g. quartz;
    (c) *even:* flat or nearly flat, *e.g.* chert;
    (d) *uneven:* rough with minute irregularities, *e.g.* quartz, barytes;
    (e) *hackly:* surface with spiky projections;
    (f) *earthy:* earthy appearance as in limonite.

Lustre	Colour	Other diagnostic features	Mineral
		Very hard (H7), lacking in cleavage, conchoidal fracture, six-sided crystals	Quartz
		Can be scratched by copper coin (H3), reacts with dilute acid, hexagonal crystals with 3 cleavages	Calcite
	Transparent, Translucent, and White Opaque	Very heavy (S.G.4–5), 2 good cleavages, massive form or tabular crystals, scratched by coin but unaffected by acid	Barytes
		Often in cubic crystals, perfect octohedral cleavage, scratched by knife (H4), sometimes colour banded	Fluorite
		Scratched by finger nail, very light in weight, may be massive, fibrous, or tabular monoclinic crystals with one good cleavage	Gypsum
Adamantine, Vitreous, Sub-Vitreous and Resinous		Salty taste, cubic crystals, good cubic cleavage, scratched by copper coin, may show colour variation	Halite
		Fairly hard (H6) only just scratched by good steel, 2 good cleavages, white, grey or bluish, twin striations	Plagioclase
		Much as plagioclase (above), but white or flesh-coloured, and no multiple twin striations	Orthoclase
		Occurs in thin colourless or silvery sheets or foliated masses due to perfect basal cleavage, flexible, soft (H2½)	Muscovite

TABLE III: CHART OF THE CHARACTERISTIC PROPERTIES OF SOME MINERALS

Lustre	Colour	Other diagnostic features	Mineral
Adamantine, Vitreous, Sub-Vitreous, and Resinous (continued)	Yellow	As described in transparent group (above)	Fluorite
		As described in transparent group (above)	Quartz
		As described in transparent group (above)	Calcite
		As described above except of slightly amber appearance	Muscovite
	Blue or Purple	As described above but can be various shades of blue	Fluorite
		As described above but of purple colouration	Quartz (Amethyst)
		Tabular crystals or botryoidal aggregates, blue streak, reacts with acid, (H3½–4)	Azurite
	Green	Olive green or yellowish-green colour, cannot be scratched by knife, 2 poor cleavages, white streak	Olivine
		Green to white foliated masses, soapy or greasy feel, very soft (H1)	Talc
		Very dark green—almost black, prismatic eight-sided crystals with 2 cleavages intersecting at 90°, hard (H5–6)	Augite
		Bright green in botryoidal masses with radiating structure, reacts with acid, pale green streak	Malachite
		As described above—usually pale blue-green crystals	Fluorite

Lustre	Colour	Other diagnostic features	Mineral
Adamantine, Vitreous, Sub-Vitreous, and Resinous (continued)	Flesh/Pink	As described above—colour may vary from pale flesh to brick pink	Orthoclase
	Pink	As described above but translucent pink colour	Quartz (Rose)
	Red-Brown	Very dark red-black colour, occurs in botryoidal masses or thin slabby crystals, fairly hard (H$5\frac{1}{2}$-$6\frac{1}{2}$), red-brown streak	Hematite
		Reddish-brown or wine-coloured crystals, frequently as dodecahedra or trapezohedra, cannot be scratched by knife (H$6\frac{1}{2}$-$7\frac{1}{2}$)	Garnet
		Adamantine crystals varying from yellowish-brown to black, good cleavage, scratched easily by knife (H4), yellow streak	Sphalerite
	Brown or Black	Brown to brownish-black, occurs in thin flexible sheets or foliated masses, perfect basal cleavage, soft (H$2\frac{1}{2}$-3)	Biotite
		As described above, but brownish-black colouration	Quartz (Smoky)
		As described above but black crystals rather than greenish-black	Augite
		Easily confused with augite it forms elongate prismatic black or greenish-black crystals, 2 marked cleavages intersection at about 60°, fairly hard (H5-6)	Hornblende
		Brownish-black adamantine crystals, very hard (H6-7), very heavy (S.G.7), pale whitish-brown streak	Cassiterite

TABLE III: CHART OF THE CHARACTERISTIC PROPERTIES OF SOME MINERALS

Lustre	Colour	Other diagnostic features	Mineral
Metallic and Sub-Metallic	Brassy or Golden	Brass-coloured, hard (H6–6½) fairly heavy (S.G.5), occurs as striated cubes or pyritohedra, or may be massive, green-black streak	Pyrite
		Golden colour with tarnished appearance, usually massive, fairly soft (H3½–4), greenish-black streak, fairly heavy (S.G.4½)	Chalcopyrite
	Silvery Grey or Black	Lead-grey or silvery grey, soft (H2½), very heavy (S.G.7½), commonly forms cubic crystals with perfect cubic cleavage	Galena
		Black or silvery black, often strongly magnetic, octahedral crystals or massive, hard (H6) fairly heavy (S.G.5)	Magnetite
		Grey or black, very soft (H1–2), greasy feel, light in weight (S.G.2), gives grey-black streak on paper, foliated masses	Graphite
		Blackish shiny crystals of flaky or slabby appearance, fairly hard (H5½–6½) and heavy (S.G.5), red or brown streak	Hematite
	Red-Brown	As above but in reddish-brown botryoidal masses or reniform (kidney-shaped) masses	Hematite

TABLE III: CHART OF THE CHARACTERISTIC PROPERTIES OF SOME MINERALS

Lustre	Colour	Other diagnostic features	Mineral
Dull or Earthy	Red-Brown	As above but in reddish-brown massive form	Hematite
	Yellow-Brown	Of variable hardness (H5–5½) and weight (S.G.2½–4), occurs mainly as dull yellow or rusty masses, sometimes with banding	Limonite
	Blue	As described above but dull, often associated with malachite	Azurite
	Green	As described above but in dull earthy masses, often associated with azurite	Malachite
Greasy	Grey-Black	As described above in metallic lustre section	Graphite
	Green-Grey or Whitish	As described above	Talc

(N.B. Whilst mineral recognition in the above chart is based initially upon the description of lustre and colour it must be emphasised that these two are often not very diagnostic or reliable properties when used alone. H = hardness,   S.G. = specific gravity.)

17. Radioactive, magnetic and electrical properties. Radioactivity is the process whereby certain invisible rays are emitted by substances, the rays being capable of penetrating bodies opaque to light. Minerals containing radium, uranium, and thorium have this property of radioactivity. Pitchblende is the most important radioactive mineral; others include monazite and thorite. Some minerals possess a magnetic property. Magnetite is affected by an ordinary bar magnet and is highly magnetic. Many others respond to the electromagnet. Most minerals containing iron are generally, though not necessarily, magnetic while there are some minerals which contain no iron which display a degree of magnetism, *e.g.* monazite. The application of friction or pressure or heat to some minerals may result in electricity being developed. Pyroelectric minerals are those which become electrically charged when subjected to heat; piezo-electric minerals those becoming electrically charged when subjected to pressure.

PROGRESS TEST 3

1. What is an element ? Name four examples of elements which exist in nature making minerals by themselves and name the four most important elements occurring in the earth's crust. **(1)**

2. Give a definition of a mineral. Which of the following are not minerals: iron, coal, halite, lead sulphide, glass, copper, diamond, guano, pyrites, cement. **(2, 4, 6)**

3. Show how mineral compounds differ from mixtures. **(3)**

4. Name the common rock-forming minerals. **(6)**

5. Name the six principal physical properties of minerals. **(7)**

6. Explain the meaning of the following terms: streak, schiller, metallic, vitreous, opaque. **(9, 10, 11)**

7. Mohs' generally adopted scale of hardness recognises ten degrees of hardness. Name these different degrees of hardness and give examples of minerals illustrative of each. **(13)**

8. "Each mineral has only one characteristic crystal form." Elaborate upon this statement. **(14)**

9. Distinguish between cleavage and fracture. Name five different kinds of fracture which are generally recognised. **(15, 16)**

10. Explain *four* of the following: perfect cleavage, sub-conchoidal fracture, specific gravity, diaphaneity, non-metallic minerals. **(4, 11, 12, 15, 16)**

CHAPTER IV

ELEMENTARY PETROLOGY

THE NATURE AND IMPORTANCE OF ROCKS

1. What is rock? Everyone is familar, in general terms, with the nature of rock and we speak of granite, sandstone and limestone, knowing that these are kinds of rock. But, to the geologist and geographer, the term rock is applied to such soft materials as muds, clays and sands as well as to hard, massive slabs or boulders of stone. Indeed, any naturally occurring agglomeration of mineral particles forms rock in the strict geological sense.

Rocks are composed of minerals, and we defined and described the character of minerals in the last chapter. As we noted there, altogether about 1,500 mineral species are known to the geologist and to these, and their varieties, many thousands of names have been given. However, only relatively few of these numerous minerals are important "rock-forming" minerals; even so, these few collectively make up something like 99 per cent of the rocks of the earth's crust.

Rocks, then, are usually composed of two or more minerals bound together and we can say that rocks are *mixtures* of their component minerals. It should be noted that, though most rocks are made up of minerals, there are a few substances of organic origin which the geologist also accepts as rocks; the chief of these are coal, peat and guano.

2. The nature of rocks. If we take a few random samples of rock, say granite, obsidian, sandstone, chalk, slate, marble, and study them, if only cursorily, we are immediately conscious of the fact that they differ very much from one another in such obvious characteristics as their colour, hardness, texture and composition. We should also strongly suspect that these different types of rocks had different origins (as, in fact, they had) and if we scrutinised them more carefully, for example by putting them under a microscope (Fig. 5), or analysed them, by for instance making chemical tests, then we should be convinced about their varying origins.

Some rocks are quite clearly mixtures of minerals, and heterogeneous rocks, such as granite, can in fact be taken to pieces and the constituent minerals separated out. Again, in the case of granite, the crystalline structure is clearly apparent to the naked eye. Obsidian, with its dark, glassy appearance,

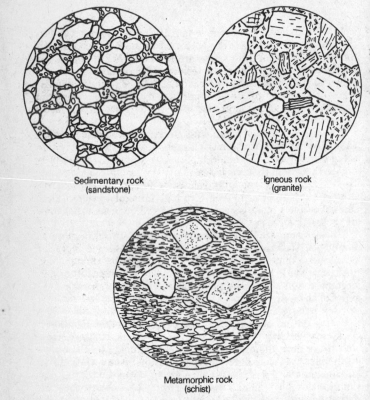

Sedimentary rock
(sandstone)

Igneous rock
(granite)

Metamorphic rock
(schist)

FIG. 5.—*Rock types*. Three examples of common rocks as seen in microscope sections.

seems to be a very different kind of rock to granite, yet they both belong to the same family group of igneous rocks. In contrast to granite, there are some rocks which are virtually made up of a single mineral as, for example, pure limestone which may consist entirely of calcite. Chalk, which is a soft, whitish, fine-grained rock of a friable nature, is basically made up of foraminifera, minute marine creatures, whereas sandstones are composed of small, individual, well-sorted grains—mainly quartz grains—cemented together.

3. The science of petrology. The whole study of geology is primarily focused upon rocks, but the special study of rocks in themselves is known as petrology. This is concerned with their origin, occurrence, composition, characteristics and classification.

There is a very great variety or rocks; some, like granite, limestone, sandstone and schist, are very common and cover extensive areas, others such as gabbro are less common and may occur in only widely separated locations.

This variety is the result of differences in the properties, sizes, shapes and arrangements of the different minerals which make up the rock.

Rocks, too, have different origins. It is possible for rocks to have the same mineral composition and yet be of different origins; for example, arkose is formed from the accumulation and compaction of transported fragments of quartz and felspar on the earth's surface while graphic granite is also composed of crystals of quartz and felspar though these crystallised out deep down within the earth's crust.

The petrologist also studies the habits of rocks; for example, some are massive in character and originated within the crust while others are stratified or were laid down in layers upon the crust. The petrologist, too, is concerned with what he terms the *primary* characters of rocks, *i.e.* those characteristics which indicate the mode of their formation, and the *secondary* characters, *i.e.* those characteristics or effects imposed upon them after their original formation; for example, a bed of peat is most likely to have been laid down originally as a more or less horizontal layer but subsequently it was turned into coal and the beds of which it was a part may have been folded; hence a folded seam of coal offers a record of the later history of an original accumulation of peat.

As Read and Watson have said:

> "A study of all the characters of a stratum of rock—its composition
> form, the arrangement of its constituents, its fossil contents and all
> that can be found out about it—leads to an opinion as to how and
> where it was made. These total characters of a rock give its facies,"
> (*Introduction to Geology*, Macmillan, 1962, pp. 5–6).

4. Methods of investigation. Petrological investigation is of
three main kinds:

    (*a*) field mapping,
    (*b*) laboratory work,
    (*c*) theoretical deduction.

Work in the field provides the fundamental basis for almost all
geological investigations; this is especially true of the study of
petrology. Field work is augmented by work in the laboratory
where experiments can be carried out and where rocks can be
studied under the microscope.

IGNEOUS ROCKS

5. The origin of igneous rocks. All igneous rocks constitute
materials that have originated from an initially molten condition. Indeed, the very name comes from the Latin word *ignis*,
meaning "fire". The molten rock is referred to as *magma* and
hence igneous rocks are also sometimes termed magmatic rocks.

Magma originates within the earth wherever temperatures
rise sufficiently high to cause melting. It is generally believed
that the heat is provided by the decay of radioactive material
but tectonic processes may also play a part. For instance,
frictional heat generated at interacting plate margins (*see* x, **8**)
may contribute a great deal to the magmatic activity occurring
in those areas. As a result of tectonic movements, crustal loading, or density differences, the magma is forced upwards through
the crust along lines of weakness until it spills out on the surface
as *lava*. This soon cools and consolidates to form *extrusive
igneous rocks*. The magma which does not find an outlet
eventually solidifies in cavities and fissures within the crust to
form *intrusive rocks*.

Using this mode of occurrence as a basis, a threefold division
of igneous rocks can be made:

(a) plutonic rocks—deep-seated and of a coarsely crystal-line character;

(b) hypabyssal rocks—intruded rocks occurring at shallow depths;

(c) volcanic rocks—extruded finely crystalline or glassy rocks.

Some add a further category to these:

(d) pyroclastic rocks—the accumulated material of explosive volcanic activity, e.g. lava clots, ash and dust.

Whilst this classification is based upon the position in which the rock finally solidifies in the crust, it should be emphasised that most magmas originate from the uppermost parts of the mantle (see VII, 7), although in certain cases fusion of the lower crust may also take place. The composition of an igneous rock, how-ever, depends not only on the composition of the original material melted, but also on the subsequent history of the magma as it passes through the crust. For example the magma might well react with or melt and incorporate (assimilate) a number of surrounding rocks during its upward passage into a magma chamber. It may also lose some of its components. Some for instance might escape as gases penetrating through the surrounding rocks; others may be lost as early formed crystals sink away through the magma because of their higher specific gravity. All such processes which lead to a progressive change in the composition of magmas are referred to as processes of *magmatic differentiation*; they are thought to be partially responsible for the great variety in igneous material found at the earth's surface.

In spite of such processes, however, there still appears to be a broad spatial pattern in the occurrence of the major magma types which is related both to the structure of the earth's crust and upper mantle and also to global tectonic systems. Thus oceanic areas, and constructive plate margins (see X, 8) in particular, are mostly zones of basic magma generation. Destructive plate margins and especially those at oceanic-continental boundaries, are characterised by the eruption of predominantly intermediate or calc-alkaline lavas. Finally acid or granitic material is chiefly found within the continental masses; it most commonly occurs as batholith intrusions in orogenic zones (see IX, 6).

6. Texture. The classification of igneous rocks into plutonic and volcanic associations places the emphasis primarily upon their mode of occurrence, but the petrologist also makes use of two other important characteristics:

(*a*) the rock texture,
(*b*) the chemical and mineralogical composition.

The texture of an igneous rock is largely determined by its rate of cooling. Slow cooling gives rise to a coarsely crystalline (*holocrystalline*) texture of a type normally found in deep-seated bodies such as batholiths; very rapid cooling, on the other hand, produces a very fine-grained (*cryptocrystalline*) or even a glassy rock, such as one finds in lava flows; those rocks which have cooled and consolidated at a moderate speed (*e.g.* in some high level intrusions such as dykes and sills) develop an inter-mediate texture. Particularly common among the latter group is texture composed of large crystals (*phenocrysts*) set in a finer-grained ground-mass. Such a texture (*porphyritic texture*) indicates that the phenocrysts were early-formed crystals whilst the ground-mass did not consolidate until a rather later phase when the magma has reached a level where cooling was much more rapid. It may be noted, however, that a porphyritic texture is not exclusively confined to hypabyssal rocks.

7. Chemical and mineralogical composition. In most igneous rocks only eight elements (oxygen, silicon, aluminium, iron, calcium, sodium, potassium and magnesium) make up virtually the total bulk of the rock—up to as much as 99 per cent; the remaining minute fraction consists of numerous trace-elements of which hydrogen, phosphorus and titanium are usually the most important.

If a chemical analysis is made of igneous rocks, the result shows that they are made of complex mixtures of a number of chemical compounds which, in their simplest form, may be regarded as the oxides of various elements. Moreover, two main groups of these oxides are readily recognisable: first, the oxide of silicon (silica) and second, the oxides of various metals (*e.g.* alumina, magnesia, etc). Analysis reveals that the former is the most abundant component, commonly accounting for between 40 and 75 per cent of the total composition. This silica percentage forms an important diagnostic character and has consequently been used as one of the bases of the classification of igneous rocks. The groups produced are as follows:

Over 66% silica	acid
52–66% silica	intermediate
44–52% silica	basic or mafic
Under 44% silica	ultrabasic or ultramafic

In practice a rock is often identified as belonging to one of these groups by its *mineralogical* rather than its chemical composition since the former can frequently be much more quickly identified. It is of course possible to differentiate between the rock groups on the basis of their mineralogy since mineral composition is very largely controlled by chemical composition. Igneous rocks are essentially aggregates of silicate minerals of which there are two major categories—the *alkali-alumino silicates* and the *ferro-magnesian silicates*. The former include the felspars and felspathoids, the latter include pyroxenes, amphiboles, micas, and olivines. The dark-coloured and relatively heavy ferro-magnesian minerals are the dominant components of basic and ultrabasic rocks whereas the lighter coloured alkali-alumino silicates predominate in acid rocks. Thus the basic and ultrabasic rocks are particularly characterised by such minerals as the olivines and pyroxenes (*e.g.* augite) whilst acidic rocks commonly contain at least ten per cent quartz and may have muscovite (mica) and a high proportion of alkali felspar. The intermediate rocks are dominated by felspars and amphiboles (*e.g.* hornblende) and may also contain appreciable amounts of pyroxenes (*see* Table IV).

NOTE: where the boxes in Table IV are divided, the rock in the upper section of the box contains dominantly orthoclase felspar; the rock in the lower section, dominantly plagioclase felspar.

8. Igneous rocks. Among the more common igneous rocks are:

(*a*) *Acid rocks.* The great bulk of the acidic rocks consists of a plutonic association which includes *granites* and *granodiorites* together with some related intermediate rocks such as *syenites* and *diorites*. These holocrystalline rocks display three very noticeable characteristics:

TABLE IV: IGNEOUS ROCKS

Silica Percentage / Geological Occurrence	VOLCANIC	HYPABYSSAL	PLUTONIC	Common Minerals
Texture	Cryptocrystalline or Glassy	Porphyritic	Holocrystalline	
		Common Rock Types		
More than 66% ACID	ALKALI RHYOLITE / RHYOLITE	GRANITE PORPHYRIES	GRANITE / GRANODIORITE	Quartz, Felspars (particularly Orthoclase), Micas, Hornblende
52–66% INTERMEDIATE	TRACHYTE / ANDESITE	SYENITE and DIORITE PORPHYRIES	SYENITE / DIORITE	Felspars, Hornblende, Micas, Some Augite
44–52% BASIC	ALKALI BASALT / BASALT	DOLERITE	ALKALI GABBRO / GABBRO	Augite, Felspar (mainly Plagioclase), Olivine
Less than 44% ULTRABASIC	Rare	Rare	PERIDOTITE	Olivine, Augite (or occasionally Hornblende)

(*i*) They are almost entirely confined to continental areas.

(*ii*) They largely occur in the form of big intrusions such as batholiths, stocks and bosses.

(*iii*) They are frequently, but not always, associated with highly metamorphosed rocks at the heart of orogenic zones.

It is thought that they result from partial fusion in the lower parts of the crust, and that the melt so produced has moved to higher levels by diapiric action and processes of magmatic stoping. Yet not all granites show such clear evidence of intrusion, and in many Pre-Cambrian terrains in particular where erosion has gone deep, the boundaries between granites and surrounding metamorphic rocks are rather diffuse and occupied by *migmatites* (rocks of mixed igneous and metamorphic parentage). Indeed there is much evidence to support the idea that some of these intrusions are of a metasomatic and not igneous origin, and that they result from the process of *granitisation*.

Acid rocks found in volcanic associations are very much less common than granites. They include *rhyolite, obsidian* and *pitchstone*, the two latter being glassy varieties. Acid volcanics are often erupted in association with andesites in orogenic zones at destructive plate margins. It is thought that they probably represent the remnant differentiate of an originally more basic lava emanating from the reprocessed crust or upper mantle. Acid lavas tend to be very viscous and frequently give rise to explosive volcanic activity and the development of *nuée ardentes* (*see* IX, **11**). Eruptions of this and related types may produce *pumice* (a sponge-like lava full of gas vesicles) and *ignimbrite* (a rock composed of many small glassy fragments or shards welded together).

(*b*) *Intermediate rocks.* As has been noted above some varieties of syenite and diorite are related to granites and may well have originated in a similar fashion. We have also seen that among the volcanic varieties *andesite* is a most abundant rock at oceanic–continental boundaries, where a lithospheric oceanic plate is subducting beneath a continental one. It has been shown in fact that there is often a variation across such an environment from tholeiitic lava on the oceanwards side through andesite of the orogenic zone to more alkaline varieties such as trachyandesite and trachybasalt on the continental side. The derivation of these lavas is still uncertain.

It was formerly thought that they might be mixtures of primary acidic and basic magmas, but it now seems more likely that they result from the partial fusion of the upper parts of a descending oceanic plate.

Some intermediate rocks can be grouped with more basic varieties in an *alkali-rich association* which is characterised by an unusually high percentage of soda and potassium. This results in their containing minerals such as felspathoids in place of felspars. *Nepheline syenite* is one such rock. The provinces of alkali-rich rocks are in continental areas; one of the best known is perhaps that of the East African domes and rift valleys. Whilst some localised occurrences of alkaline rocks may be the result of limestone contamination of a magma, in such large areas as Eastern Africa it is now believed that both the magmas and the domes are connected with rising plumes in the asthenosphere.

(c) *Basic rocks.* In striking contrast to the acidic group, the great bulk of basic rock occurs in the form of lavas or high level intrusions. Basalt makes up over 98 per cent of all volcanic rocks and forms the crust of all oceanic areas as well as appearing in many continental regions. There are two major types of basalt, *tholeiite* and *alkali olivine basalt*. The former has a higher silica content than the latter and may actually contain a little quartz. Tholeiites are particularly generated at mid-oceanic ridges (constructive plate margins) but the basalt magma becomes increasingly alkalic with distance from this environment. Both types appear to be produced from the partial fusion of the earth's peridotite mantle. *Dolerite* is the hypabyssal equivalent of basalt and is one of the chief rocks forming dykes and sills (*see* IX, 6). The plutonic representative, *gabbro*, is perhaps surprisingly much less common than basalt. Where it does occur (*e.g.* the Bushveld lopolith) it often forms banded or layered intrusions, a feature produced by the gravity settling of minerals.

(d) *Ultrabasic rocks.* Although small amounts of ultrabasic material are produced as differentiates of basic magmas, the great proportion of such rocks are found as large intrusions in orogenic zones (*e.g.* Circum-Pacific belt) where it is thought that they represent mantle material injected into sediments in the regions of deep ocean trenches. The main rock types found are *peridotite* and *serpentinite*. One major problem associated with these is that although they have very high melting points the surrounding rocks show little sign of contact metamorphism. It has therefore been argued that they may well have been emplaced by tectonic action whilst in a solid or near solid state.

SEDIMENTARY ROCKS

9. The origin of sedimentary rocks. These are "derived" rocks which at the time of their original formation were laid down either under water or on land surfaces in most cases as more or less horizontal layers. They are mainly the result of surface processes which have produced the raw materials of which they are composed, although some have originated through other processes, *e.g.* some organic (or biogenic) rocks. No classification is entirely satisfactory but, broadly speaking, it is possible to distinguish two main groups (*see* Table V):

(*a*) *clastic or detrital sediments, i.e.* those made up of fragments of mineral or rock matter; such fragments have usually been transported in some way before being laid down;

(*b*) *chemical and organic sediments, i.e.* those which result from chemical precipitation, often in evaporating seas, and those which are produced by biotic activity.

The first of these groups is commonly subdivided on the basis of grain size into:

(*i*) *Rudaceous sediments.* In these, 50 per cent or more of the rock fragments are fairly large, *i.e.* over 2 mm. in diameter. If the particles are rounded the rock is termed a *conglomerate,* if angular a *breccia.* Conglomeratic material develops on beaches and in rivers, especially where the latter discharge from mountainous areas on to plains. They also sometimes form when a marine transgression (*see* glossary) takes place across a denuded land surface. Forming at a plane of unconformity, such deposits are termed *basal conglomerates.* Breccias and conglomerates are also occasionally found in rather deeper water marine environments as a result of transportation in submarine slumps. Beds of this origin may be referred to as *slide*

TABLE V: SEDIMENTARY ROCKS

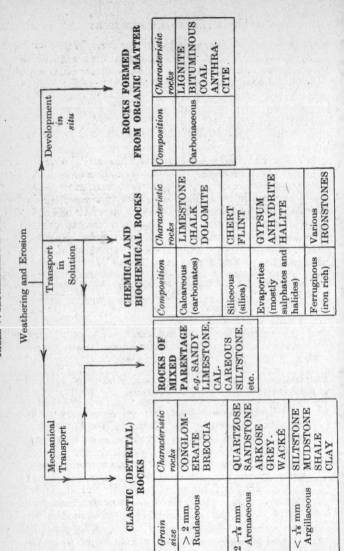

Weathering and Erosion

Mechanical Transport · Transport in Solution · Development *in situ*

CLASTIC (DETRITAL) ROCKS

Grain size	Characteristic rocks
> 2 mm Rudaceous	CONGLOMERATE BRECCIA
2 – 1/16 mm Arenaceous	QUARTZOSE SANDSTONE ARKOSE GREYWACKÉ
< 1/16 mm Argillaceous	SILTSTONE MUDSTONE SHALE CLAY

ROCKS OF MIXED PARENTAGE e.g. SANDY LIMESTONE, CALCAREOUS SILTSTONE, etc.

CHEMICAL AND BIOCHEMICAL ROCKS

Composition	Characteristic rocks
Calcareous (carbonates)	LIMESTONE CHALK DOLOMITE
Siliceous (silica)	CHERT FLINT
Evaporites (mostly sulphates and halides)	GYPSUM ANHYDRITE HALITE
Ferruginous (iron rich)	Various IRONSTONES

ROCKS FORMED FROM ORGANIC MATTER

Composition	Characteristic rocks
Carbonaceous	LIGNITE BITUMINOUS COAL ANTHRACITE

conglomerates or *breccias*. Finally, many glacial deposits are of a very coarse-grained nature. Indurated glacial deposits are called *tillites*.

(*ii*) *Arenaceous sediments*. These include sediments in which most of the particles have diameters of between 2 mm. and 1/16 mm. They include three principal families. The first of these, *arkoses*, are highly felspathic sandstones derived from the disintegration of granitic rocks. They may occur as thin weathered blankets covering a granitic rock or as much thicker wedge-shaped deposits that are associated with the rapid fluvial erosion of a granitic massif.

The second family are the *greywackés*. This name, which is derived from the German term *grauwacke* (grey sandstone), is applied to sandstones which contain a high proportion of argillaceous matrix and thus often tend to be dark-coloured. Greywackés are particularly common sediments in Palaeozoic rocks and it seems possible that much of the clay matrix could have been derived from the decomposition of some of the larger mineral and rock grains into secondary chloritic material. Most, although not all greywacké beds exhibit structures such as graded bedding and flute casts suggesting that they represent the deposits of turbidity currents; the latter are turbulent density flows of sand and mud normally generated by sub-aqueous sediment slides on unstable slopes.

The final family are the *quartzose sandstones*. As the name suggests, these are well sorted sandstones composed very largely of quartz grains. They normally display structures such as cross-stratification and ripple marking indicating a shallow water origin of either a fluvial or marine nature.

(*iii*) *Argillaceous sediments*. This group of sediments have grains less than one sixteenth of a millimetre in diameter. They are largely composed of very fine silica particles and clay minerals. When indurated, the coarser grained varieties are referred to as *siltstones*, the finer as *mudstones*. Flakey clay minerals, however, frequently become orientated parallel to the bedding and so impart a fissility to the rock. Argillaceous strata exhibiting such bedding fissility are described as *shales*. Muds occur in a number of colours and accumulate in a variety of environments—in lakes, lagoons, deltas, and shallow and deep seas.

In each case currents must be insufficiently strong to carry in coarser terrigenous detritus.

The second major category of sediments, the chemical and organic sediments, are usually classified initially by their chemical composition rather than grain size. Yet it should be emphasised that a great many rocks are in fact a mixture of both clastic and chemical components. Among the more important groups of chemical and organic rocks are:

(i) *Limestones*. These are a series of rocks composed mainly of calcium and magnesium carbonates. Their material is retrieved from aqueous solution either by chemical precipitation or as a result of organic extraction. Organic limestones often contain diagnostic structures such as algal banding or various types of reef formation, but it is quite common to find the material reworked to produce clastic limestones, (*e.g.* calcirudites, calcarenites, calcilutites). The latter may contain such structures as cross-stratification indicating that transport has taken place. There are many well known varieties of limestone. *Oolitic limestone* is a type composed of numerous small spheroidal ooliths that form from gentle current activity and chemical precipitation in warm shallow seas (*e.g.* on the Bahamas Banks). *Chalk* is a particularly pure and very fine grained, porous limestone which is almost wholly composed of shell debris. The debris consists of fragments of microfossils such as foraminifera together with the remains of planktonic algae. A limestone which contains over 50 per cent of the mineral dolomite (the double carbonate of calcium and magnesium) is termed a *dolomitic limestone*. It may result from the precipitation of dolomite in an evaporating sea, or perhaps more probably from the metasomatic alteration of normal limestone deposits by magnesium rich brines.

(ii) *Evaporite deposits*. A large number of salts may originate from the dehydration of sea or lake waters. Particularly common are the chlorides and sulphates of such elements as sodium, calcium, potassium and magnesium. In view of the fact that the evaporation of 300 metres of sea water would provide less than 5 metres of salts, it seems surprising that many ancient salt deposits are over one hundred metres thick. It is generally believed that they may have been formed in one of two ways. Firstly they may have accumulated in evaporating and gently subsiding barred sea basins which were periodically being recharged with water. A second and perhaps more likely hypothesis is that they resulted from precipitation in zones of marginal salt flats which were only occasionally inundated by the sea. Such an environment can be seen at the present time in the Mexican salt flats of the Ojo de Liebre.

(*iii*) *Ferruginous deposits*. Deposits having a sufficiently high content of iron to allow them to be economically worked are often loosely referred to as ironstones. They occur as sediments in four forms—sulphides, carbonates, oxides and silicates. Some of these appear as primary deposits and others as replacement deposits, although the exact conditions under which they were produced are often not entirely known.

(*iv*) *Siliceous deposits*. The best known deposits of this type are *chert* and *flint*. Formed of cryptocrystalline silica, they differ in that flint possesses a perfect conchoidal fracture whereas chert breaks to produce a flat or rather uneven fracture. They are commonly found associated with limestones; in Britain chert is typically found in the Carboniferous Limestones and flint in the Chalk. There are many occurrences of bedded chert and flint suggesting that some form of primary deposition, perhaps from silica gels, must have taken place in marine environments. On the other hand there is also clear evidence that many nodules of silica are of a secondary or replacement origin.

Other types of siliceous deposit included *diatomaceous* and *radiolarian earths* which are composed of the silica skeletons of minute algae and protozoans (*see* VI).

(*v*) *Carbonaceous deposits*. Incomplete oxidation of dead vegetation in wet acid conditions may lead to the formation of peat deposits. As such deposits become buried beneath other sediments, so increasing pressure and temperature causes a loss of volatiles and an increasing degree of coalification. The rocks produced in this way are termed the *Coal Series* and are classified according to rank. The higher the rank the greater is the carbon content of the coal. *Lignite* and *Brown Coal* are low rank varieties in which much of the original woody structure is preserved. *Bituminous Coal*, which is of higher rank, has a characteristic banding or striping consisting of alternating dull grey (*durain*) and shiny black (*vitrain*) layers. The highest rank coal is *anthracite* in which scarcely any of the original vegetable structure can be detected. Thought to have formed only in zones of unusually high stress, anthracite possesses an almost sub-metallic lustre and a very distinctive conchoidal or sub-conchoidal fracture.

10. Textures. Texture is concerned with the size and shape of grains, their sorting and arrangement, and the physical properties of sedimentary rocks, such as porosity and permeability, which result from their textural character. As we have noted above, grain size is used as a basis for the classification of clastic sediments. Grain shapes can be described as being angular, sub-angular, or rounded and can provide a clue as to origin; for example, angular grains are unlikely to have suffered much transportation, whilst rounded grains have been much abraded and may well have been wind-borne. Grains are bound together either by a *cement* which results from chemical action or by a *matrix* which is of clastic origin.

11. Sedimentary structures. Sedimentary rocks exhibit a wide variety of structures, some of which are clearly of depositional rather than tectonic origin. Such structures are produced either at the time of formation or shortly after the formation of the sediment. They have proved invaluable to the geologist in a number of ways. In particular they are of considerable use in indicating the kind of conditions under which a sediment is laid down (*e.g.* rain pits and dessication cracks reveal that a sediment was exposed above water level). Among the more common sedimentary structures are:

(*a*) *Stratification.* Bedding and bedding-planes are the most important structural features of sedimentary rocks. A *bed* is a sheet of sedimentary material which is of great areal extent in relation to its thickness; it is the outcome of a single act of sedimentation. Each bed is separated from its neighbour by a perceptible break caused by a change or pause in sedimentation. Such a plane of separation is termed a *bedding-plane.* Within any bed a number of *laminations* or slightly varying layers may often be discerned: such laminations are usually the result of slight changes in the supply and character of the sediment. Through a prolonged process of sedimentation a succession or series of beds is built up to give a stratification.

(b) *Graded bedding.* Some beds display a vertical gradation in grain size. Normally the grading is from large grains at the base of a bed to smaller ones at the top. It is a common structure in greywacké sandstones; here it results from deposition from turbidity currents.

(c) *Cross-stratification.* This is a very common structure (*see* Fig. 6) in sandstones. There are many variants but most

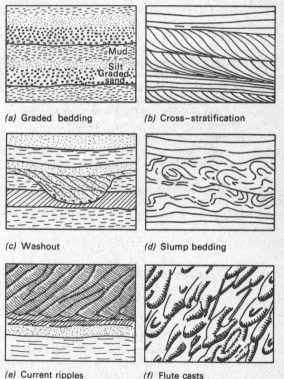

(a) Graded bedding

(b) Cross-stratification

(c) Washout

(d) Slump bedding

(e) Current ripples

(f) Flute casts

Fig. 6.— *Sedimentary structures.*

are formed as the result of slipping or avalanching of sand down the lee slope of ripples, dunes, sand-bars, fans, small deltas, etc. Cross-stratification is useful both as an indicator of palaeocurrent directions and also as a "way-up" pointer in sequences that have been greatly folded.

(d) *Washouts*. Rivers, streams, and submarine currents may erode channels in sediments. Should these channels become infilled with later deposits and preserved in a stratal sequence, they are termed washouts. They are particulary common in deltaic sequences.

(e) *Slump structures*. A great variety of folded and crumpled structures results from the downhill submarine sliding of sediments. If the direction of movement can be ascertained, some indication of the palaeoslopes may be gained.

(f) *Ripple marks*. Various ripple patterns are developed in sediment as a result of marine, fluvial, or wind currents. Like cross-stratification they reveal palaeocurrent directions.

(g) *Sole markings*. In some interbedded sequences of sandstone and shale (particularly greywacké sequences), a variety of sedimentary markings may occur on the bases of the sandstone units. These are collectively referred to as sole markings. Very often they are casts of structures formed in the underlying mudstone. *Flute casts* are an example of this; hollows or flutes are eroded in the mud of the sea floor by turbidity currents and then infilled and covered with sand. When the sediment is indurated and exposed at the surface, the sand infillings stand out clearly as casts on the base of the sandstone bed (Fig. 6). Many organic structures may be preserved in a similar way (*e.g.* worm tracks).

12. Deposition of sediments. The greatest theatre of sedimentation is the sea. The general tendency is for the land waste, resulting from the decay and disintegration of the rocks, to be carried by the rivers to the sea. The bulk of the sediment is deposited in the shallow water on continental margins. Such land-derived material is reinforced by large quantities of material worn from the coasts by marine erosion. Over vast areas of the ocean floor shells and skeletal remains of animals, as well as volcanic dust, accumulate, though much more slowly. Most lake bottoms receive sediments carried in by streams and so lacustrine deposits accumulate. Ice-sheets may leave behind thick spreads of glacial moraine. Also, mineral matter in solution, such as halite and gypsum, may be precipitated as a result of evaporation. Again, vegetable matter, which eventually changes into peat, lignite and coal, accumulates in swamps, bogs and some lakes.

While the bulk of sediments are *aqueous*, *i.e.* laid down in water, various types of sediments are deposited directly upon the land, for example:

(*a*) piles of rock fragments derived from cliff faces or steep slopes accumulate at their bases;

(*b*) wind, especially in arid regions, transports and deposits vast quantities of dust and sand;

(*c*) glaciers and ice-sheets carry and lay down large amounts of rock waste directly upon the land;

(*d*) mineral-charged waters (springs) emerging from the earth often deposit their mineral matter at the surface.

13. Consolidation of sediments. Most sedimentary deposits are consolidated into relatively hard rocks, though when they were originally deposited they were mostly loose, incoherent masses of material, *e.g.* sandstone was once loose sand, and shale soft mud. The process of converting sediments into sedimentary rocks is termed *lithification*. The various processes leading to lithification are known as diagenesis. Diagenesis may be defined as comprising all those changes that take place in a sediment near the earth's surface at low temperature and pressure and without crustal movement being directly involved (Read & Watson, *Introduction to Geology* p. 255). The diagenetic processes are twofold:

(*a*) *physical* which include dewatering, compaction and welding;

(*b*) *chemical* which embrace desalting, recrystallisation, cementation and replacement.

These operations may occur conjointly.
Two of the most important processes in the consolidation of sediments are:

(*a*) downward pressure or weight: where strata pile up in great thicknesses, the weight of the overlying mass expels water and squeezes the fragments of the lower layers to-

gether causing them to consolidate, perhaps by adhesion and
cohesion.

(b) cementation: water laid down with the rock particles
or penetrating the detrital matter may carry various minerals
in solution; various cementing minerals may be deposited in
the interstices of the sedimentary deposit and cause the
whole mass to become tightly bound together.

It should be noted that the diagenetic process, with increasing
pressure and temperature at depth, passes into metamorphism.

METAMORPHIC ROCKS

14. Metamorphism. The term metamorphism comes from
the Greek *meta-morphe*, which means "change of form"; hence,
metamorphic rocks are those which have been altered and show
changes in character and appearance. Since, however, rocks
remain essentially solid during the process of metamorphism
they retain, usually, some of the primary characteristics be-
queathed to them from the parent material; but in their altered
form their composition and structural and textural features are
determined in part by the character of the parent material and
in part by the conditions of metamorphism.

Metamorphic rocks were originally either igneous or sedimentary rocks. Although it is often possible to tell what the
original rock was, sometimes the change has been so great that
it is impossible to say with any certainty from which rocks the
metamorphosed rocks were originally formed; some gneisses
provide a case in point.

All rocks can be changed to a greater or lesser extent, but
the ease with which they are changed and the degree of alteration is related to several factors, the most important of which
are:

(a) the resistance of the rocks to pressure;

(b) the size of the constituent grains of the rocks;

(c) the degree of porosity of the rocks;

(d) the solubility of the constituents of the rocks;

(e) the chemical reactions of the minerals in the rocks;

(f) the stability of the mineral assemblage that is produced.

15. Causes of metamorphism. The change in the character

of rocks from their original state may result from a variety of actions and factors:

(a) *Temperature increase* may induce recrystallisation. Rocks may be subjected to heating either as a result of contact with a hot magma or by deep burial within the crust where much higher temperatures prevail.

(b) *Pressure increases* also produce metamorphism. The most important stress acting upon rock is hydrostatic stress or the weight of the overlying strata. Tectonic stresses may act upon rocks to induce metamorphism but such stresses usually act for relatively short periods.

(c) *Pore-fluid pressure* has an important bearing upon metamorphism. In porous rocks interstitial fluid may be important in two ways. Firstly it increases pressure and secondly it acts as a catalyst and may well be involved in metamorphic reactions. It should be emphasised in this connection that metamorphic changes do not involve any changes in the bulk chemical composition of the rock. Nothing is introduced into the rock or leaves it with the exception of volatiles such as water or carbon dioxide. Where changes in chemical composition do occur as the result of the introduction of new elements by percolating fluids or vapours, the process is referred to as *metasomatism*.

(d) *The time factor* is also important in metamorphism. A rock subjected to a particular stress for a short period may develop a different mineral assemblage from a similar one subjected to the same stress for a much longer time.

16. Metamorphic textures and structures. As has already been stated, the textures and structures of metamorphic rocks depend partially on the nature of the original rock and partially upon the metamorphic processes which produced them. Whilst metamorphism has an overall tendency to coarsen textures, it may generally be said that low temperatures and high pressures favour fine-grained textures and high temperatures and prolonged metamorphism favour coarser textures. The main textures found in metamorphic rocks are:

(a) *Cataclastic texture*. A texture produced as the result of the fracturing and warping of minerals under tectonic stress.

(b) *Hornfelsic texture*. A texture in which minerals grow in a randomly orientated fashion so producing a rather massive rock with a very irregular fracture.

(c) *Granular texture*. A granular form occurs when minerals develop in an equidimensional way to produce a welded mosaic of crystals. Rocks such as marble and quartzite frequently exhibit granular textures.

(d) *Slaty texture*. High stresses applied to argillaceous rocks produce a strong parallelism of the clay minerals and impart a strong slaty cleavage.

(e) *Phyllitic texture*. This is similar to but slightly coarser than a slaty texture. The development of mica along parallel planes produces a planar orientation in the rock known as *foliation*.

(f) *Schistose texture*. Schistosity is very strongly foliated texture produced by the growth of minerals such as chlorite, mica, and hornblende.

(g) *Gneissose texture*. A very coarsely crystalline texture in which a type of foliation is produced by the segregation of minerals into lighter and darker bands.

Structures in metamorphic rocks also often depend upon the characters of the original rock. Thus features such as stratification and even graded and cross-bedding can still be recognised in some metamorphic sequences. The most strongly metamorphosed zones, however, are also often found to be zones of very severe tectonic deformation (*e.g.* the North West Highlands of Scotland). In such terrains there is much evidence to suggest that the rocks frequently attained a state of plasticity during deformation. The folds therefore show little of the regular geometry of flexures in brittle rocks, and the faults most commonly take the form of drawn out shear planes. It is in such environments at the heart of orogenic belts that we find metamorphism associated with large scale thrusting and nappe formation (*see* VIII).

17. Types of metamorphism and associated rocks. Metamorphic rocks are named by their texture and mineral assemblage. They can be classified into groups identified as belonging to a particular *metamorphic facies* (*e.g.* greenschist facies, amphibolite facies, etc.). A metamorphic facies may be defined as a collection of rocks which has been formed under the same broad conditions of temperature and pressure. Thus it is possible for rocks of a very different appearance to belong to the same facies. It is also possible to classify metamorphic rocks as being associated with a particular type of metamorphism, although such a subdivision is not entirely satisfactory. The different categories of metamorphism include:

(*a*) *Contact or thermal metamorphism*. Contact metamorphism takes place in conditions of elevated temperatures but where pressure increases play no significant part. Such conditions are provided by contact with hot magma. The grade of metamorphism depends not only upon the temperature of the igneous body but also upon its size. Small intrusions such as dykes and sills may only produce a slight amount of marginal "baking", but a large granite pluton produces a zone of alteration (the *metamorphic aureole*) which may extend for well over a kilometre. The greatest alteration occurs nearest the intrusion where metasomatic action is also common. Rocks which are commonly produced by contact metamorphism are:

(*i*) *Hornfels*. A dark, tough, granular rock containing poorly developed metamorphic minerals such as andalusite and cordierite.

(*ii*) *Spotted slate*. A highly cleaved argillaceous rock characterised by scattered small nodules or "spots". The latter are in fact very impure crystals of andalusite and cordierite.

(*iii*) *Marble*. A rock of sugary or granular texture produced by the metamorphism of limestone.

(*iv*) *Quartzite*. Quartzites are very hard granular rocks produced by the recrystallisation of sandstones. Both marbles and quartzites are also produced by other types of metamorphism although a more lineated rock fabric is produced when stress is involved.

(*b*) *Dynamic metamorphism*. This is due to the action of great pressure causing dislocation, crushing and shearing. Its effects are best seen in great fault and crush zones. Although dynamic metamorphism is brought about by stress, heating produced by frictional effects may also play a part in localised areas. Among the rocks produced by this sort of metamorphism are:

(*i*) *Cataclastic breccia*. A material produced by the mechanical break up of the rocks. It is often composed of angular fragments set in a finer matrix.

(*ii*) *Mylonite*. A microbreccia composed of very minute angular particles bound together by an indurated rock powder. As such a material is sheared further it may develop a striping or banding.

(*c*) *Regional metamorphism*. Metamorphism of this sort is a much larger scale phenomenon than the previous types and does in fact cover vast areas. Regionally metamorphosed rocks occur in orogenic belts and are therefore associated with strong deformation. The metamorphism varies in grade from a low-grade variety where rocks show little difference from diagenetically altered forms, to a high-grade association which terminates in partial melting. There does not appear to be, however, any simple relationship to depth, since the intensity of metamorphism has been shown to vary in a complex way laterally as well as vertically. It has proved possible to map the grade of metamorphism by establishing a series of *zones* of progressive metamorphism, each of which is characterised by the formation of a particular mineral—the *index mineral*. The pioneer work in the recognition of such zones was achieved as long ago as 1893 by George Barrow in his classic study of the rocks of the south-eastern part of the Scottish Highlands. Typical rocks occurring in areas of regional metamorphism include:

(*i*) *Slate*. Located in zones of low grade metamorphism, a slate is a very highly cleaved, argillaceous rock.

(*ii*) *Phyllite*. A slightly coarser grained rock than a slate, it has a silky appearance due to the covering of fine micas along the planes of parting.

(*iii*) *Schist*. The rock which is most often associated with regional metamorphism, this is produced in medium-grade metamorphic environments. Its most characteristic feature is its foliation which is produced by the parallel growth of minerals such as mica, chlorite, and hornblende.

(*iv*) *Gneiss*. Produced by the highest grades of metamorphism, a gneiss is a coarsely crystalline rock of a granular texture. It often contains alternating bands or lenses of felspar and ferromagnesian minerals.

TABLE VI: METAMORPHIC ROCKS

Type of Metamorphism	Geological Occurrence	Processes	Texture	Characteristic Rocks
CONTACT or THERMAL	Adjacent to igneous rocks, and particularly near large granite intrusions	Recrystallisation and metasomatism as a result of rising temperature and reaction with invading fluids and gases	Predominantly granular textures	Spotted slate Hornfels Marble Quartzite
DYNAMIC	In intensely crushed zones, particularly around large thrust and wrench faults	Brecciation, crushing and shearing	Cataclastic textures	Breccia Mylonite
REGIONAL	Associated with orogenic belts	Recrystallisation under a variety of pressure–temperature conditions	Mainly strongly foliated or banded textures	Slate Phyllite Schist Gneiss

ROCKS AND THE LANDSCAPE

18. The importance of rocks. In the previous two chapters attention has been directed to the nature and characteristics of rocks and to their origins and classification. Before moving on to other aspects of geology, it may not be inappropriate to consider, if only briefly, the importance of rocks generally and, more particularly, their role in landscape formation.

In general terms, rocks are important because they influence:

    (a) the general character of the landscape;
    (b) the development of particular landforms;
    (c) the nature and qualities of soils;
    (d) the occurrence of mineral wealth;
    (e) the availability of building materials;
    (f) the quality and quantity of surface and sub-surface water supplies.

Soils, for example, though they result from a variety of processes and conditions, are fundamentally derived from rock waste which provides the skeletal mineral matter from which mature soils are developed. A particular rock type *may* give a soil distinctive characteristics and qualities. Again, the quality and quantity of water supplies are closely related to rock characteristics such as their porosity, permeability, water-holding capacity, chemical character, etc. However, more will be said about the importance of rocks to man in the chapter on Applied Geology.

19. The influence of rocks. Although the processes of erosion depend basically upon climate, rocks may strongly affect the nature of the topography and help to determine the character of specific landforms through their composition, texture, and structure. The fact that rocks may exert a strong influence upon the landscape can be seen by comparing a topographical map (which delineates the relief features) with a geological map (which shows differing rock types), for frequently there is a close correlation between the two maps of the same area; in other words, the surface relief may be a reflection of the underlying rocks. For example, in a broad way the highest parts of the British Isles in the north and west correspond with the areas of

older, harder rocks, the lower lying parts in the south and east
with the areas of younger, softer rocks.

Although, in general terms, what has just been said may be
deemed to be true, caution must be exercised and a purely
determinist approach avoided. There is a definite tendency for
certain rocks of a hard, tough, resistant character, such as
granites, gabbros, quartzites and some kinds of sandstones and
limestones, to form high mountain and hill areas, while other
rocks which are rather soft and much less resistant in character,
such as shales and clays, are apt to give rise to vales, lowlands
and plains. Nevertheless, such generalisations "cannot always
be sustained, for erosional history and climatic influences often
modify the expectable relationship between rock-type and
landforms." (R. J. Small, *The Study of Landforms*, Cambridge
University Press 1970, p. 114.) One could quote many examples
to support this statement. For example, the highly durable
Lewisian gneiss of Northwestern Scotland and the Hebrides
forms essentially low-lying terrain. Again, one could give examples where differing rock outcrops give rise to no very perceptible
relief differences, as in the case of the Oxford, Ampthill and
Kimmeridge Clays of the Upper Jurassic series in the East
Midlands of England which show no distinguishable relief
differences, the three together forming a broad belt of generally
flat, low-lying land.

20. Lithological characteristics. The important lithological
features which affect landform development are the chemical
and mineralogical composition, and the texture.

(a) *Chemical and mineralogical composition*. The mineralogical make-up of a rock largely determines its resistance to
the agencies of denudation. The chemical composition
controls the degree to which the rock succumbs to weathering
processes whilst mineralogical hardness influences its
susceptibility to abrasive erosional actions. Certain types of
rocks as a result of their particular chemical composition are
especially prone to decay and solution. The most notable is
limestone which is especially prone to the processes of carbonation and corrosion since the calcium carbonate, of which it is
largely composed, readily reacts with mildly acidic water
(*see* XI, 8). There are also many other rocks, such as sandstones having a calcareous or ferruginous cement, which

crumble under the attack of acidic solutions. Many igneous
and metamorphic rocks are also subject to chemical decay.
The ferro-magnesian minerals and felspars which they contain are susceptible to chemical weathering particularly in
warm, humid conditions. Yet in areas such as Britain
igneous rocks generally tend to be more resistant than surrounding sedimentary rocks. Thus, for example, intrusions such
as stocks and bosses are apt to yield upland masses with
radial drainage systems whilst features such as dykes
produce linear ridges or in extreme cases wall-like landforms.
Igneous landforms of many types are well seen in Arran (*see*
Fig. 7).

The mechanical strength of a rock or its resistance to
abrasion depends upon the hardness of its constituent minerals
and the way in which they are cemented together. It is also
affected by planes of structural weakness such as jointing and
cleavage. The mechanical strength of a rock retards such
erosional processes as glacial abrasion and corrosion by
rivers. Rock hardness may also predispose towards steepness
of slope. For instance, the great strength and resistance to
shearing of the dolomitic limestone of the Italian Dolomites
has given rise to vertical cliff faces which are an outstanding
feature of the area. Conversely in areas of clay and unconsolidated sands vertical or steep faces are fairly rare since such
soft rocks can seldom sustain a very high angle of slope.

(*b*) *Texture.* The texture of a rock depends upon the sizes
and shapes of its grains or crystals along with the way in
which they are packed and cemented together. It has an
influence on the resistance of rock to denudaton since a
property such as *porosity* is very much related to it. Porosity
refers to the pores or voids in a rock. The more voids there are
the easier usually will it be for water to percolate into the
rock. The chemical actions and mechanical action (on
freezing) of such interstitial water may quickly lead to
granular disintegration. In a very general way coarse grained
rocks tend to be more susceptible to such disintegration than
fine grained ones.

21. Structural characteristics. All rocks contain planes of
structural weakness. Some of these, such as bedding planes in
sediments and tensional joints in igneous rocks, result from
processes operating during or shortly after their formation.

Others, such as planes of cleavage and faulting, owe their origin to tectonic events occurring long after the formation of the rock. Such structural planes facilitate the passage of water into and through rocks (*i.e.* increase their permeability) and so help the processes of mechanical and chemical breakdown. Structural features are of great importance in landscape development. They commonly influence the general pattern of surface relief since drainage lines tend to become incised along them. Thus lines of folding and faulting may determine the trends of ridges and valleys. Structure also affects landform in many smaller ways. Jointing, for example, is frequently instrumental in the shaping of cliff profiles and bedding sometimes determines surface slopes. This is clearly seen in the gently dipping Carboniferous Limestone zones of the Pennines north of Malham and Settle where areas of plateau-like *limestone-pavement* are developed along bedding planes and a very marked series of joints are weathered into gaping fissures or *grykes*. These same joints largely control the profiles of the many rock cliffs or scars that are also characteristic of the area.

22. Landforms and rocks. We have seen above that the lithological and structural properties of a rock can strongly influence the topographical development of a area. In some instances they become so important a control on landscape that geomorphologists refer to the scenery as being characteristic of the rock type. Thus terms such a "karst landscape", "chalk landscape", "gritstone landscape", "granite landscape", etc. are often used.

Karst landscapes developed on non-porous and well jointed limestones are highly distinctive and contain many characteristic features, *e.g.* scars, defiles, pavements, sink-holes and solution hollows, dry valleys, underground caverns, and collapse gorges. Once again it should be noted, however, that karst landforms developed under tropical conditions are not entirely similar to those of the mid-latitudes.

The Chalk landscapes in England are often of very distinctive appearance. They are typically formed of low, rolling, gently-rounded hills cut by frequent dry valleys; the lack of surface drainage on much of this porous rock helps to maintain it as a distinctive landscape of moderately upstanding character, *e.g.* the Yorkshire Wolds and the Downs of Kent and Sussex, notwithstanding the relative softness of some of the Chalk.

Gritstones and sandstones may also produce some notable landforms, particularly where they are interbedded with less resistant rocks such as shales. Tilted strata of this kind may give rise to cuesta features composed of bold escarpments and gentle dip-slopes leading down to intervening vales (*see* Fig. 7). The

(a) SECTION FROM MALVERN TO CHILTERN HILLS SHOWING HILL RIDGES AND VALES

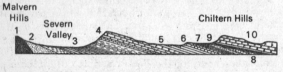

1. Precambrian and Cambrian
2. Triassic
3. Lias
4. Lower Oolites
5. Oxford Clay
6. Corallian
7. Kimmeridge Clay
8. Portland Beds
9. Gault & Upper Greensand
10. Chalk

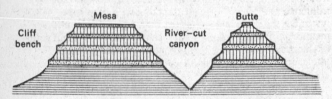

(b) SECTION ACROSS LEVEL-BEDDED SEDIMENARY ROCKS IN ARID COUNTRY

FIG. 7.—*Landforms and rocks.* (*a*) shows the effect of tilted strata of differing resistances to erosion; the result is the production of "scarp and vale" topography. (*b*) shows the effect of alternating level-bedded sedimentary rocks where there is a resistant capping rock; this kind of landscape is common in arid lands and produces a "badland" landscape, *e.g.* in Arizona.

scenery produced by the Millstone Grit Series in the West Riding of Yorkshire is very much of this type. Where strata are horizontally bedded, plateau surfaces may be produced; the old plateau of Brazil for example is capped by thick beds of sandstone which are dissected by rivers. The extensive level tablelands so produced are called *tableiros*.

More massive rocks such as granite give different effects. They frequently produce upland or mountainous scenery sometimes with a great deal of rock exposure. Under such circumstances the jointing exerts a very strong influence on the scenery. The characteristic jointing of granite rocks is well seen in the *mural* jointing of mountains such as Goat Fell on the Isle of Arran and in the residual granite blocks or *tors* of Dartmoor. Yet under more tropical conditions weathered granites and gneisses may produce much more rounded forms *e.g.* the distinctive "sugarloaf" hills in the neighbourhood of Rio de Janeiro.

We have seen then that within zones of distinctive climatic conditions, rocks may be largely instrumental in determining the physical character of a region. Perhaps this is nowhere better seen than in the British Isles where the rapid variation in geology leads to an immense diversity of scenery for so small an area.

PROGRESS TEST 4

1. What are the aims and methods of petrology? (3, 4)
2. What factors are used in the classification of igneous rocks? (6, 7)
3. What are the origins of basic and ultrabasic igneous rocks? (8)
4. Describe the nature and origin of conglomerate, greywacké and arkose. (9)
5. Give an account of the variation in limestones. (9)
6. With reference to examples show how sedimentary structures can be of use to the geologist. (11)
7. Outline the causes of metamorphism and the factors influencing its action. (15)
8. What do you understand by metamorphic aureole, foliation, metasomatism and gneiss? (15, 17)
9. Describe with examples the lithological characteristics affecting landform development. (20)
10. Outline the characteristic features of karst and chalk scenery. (22)

CHAPTER V

FOSSILS AND THE PRINCIPLES OF PALAEONTOLOGY

FOSSILS

1. What is a fossil? The word "fossil" is derived from the Latin "fossilis," meaning "dug up." Geologists use it to mean any trace of former life which appears in the rocks. Thus a fossil may be the following:

(*a*) The actual organism itself; it is extremely rare to find an animal preserved in entirety, but under certain conditions it may occur, *e.g.* mammoths found preserved in permafrost areas.

(*b*) The shell or bony structure of a creature; the hard parts of animals often decay less readily than the flesh and thus are more likely to be preserved. This is particularly true of groups having chitinous or phosphatic skeletons.

(*c*) The petrified or mineralised remains of an animal; in most fossils the hard parts of the former animal become replaced by various minerals (particularly common are silica, calcite and iron pyrites) giving a reasonably exact reproduction of the original.

(*d*) A carbon impression of an organism which has decayed away; this is a particularly important mode of preservation of plants and of soft-bodied animals such as jellyfish. As the organism breaks down after death so the volatile constituents may escape leaving behind a thin film of carbon.

(*e*) A mould or cast; these can be formed in many ways— insects trapped in pine resin, animals trapped in volcanic lavas and ashes, and footprints and worm tracks, etc. Casts represent the infilling of moulds and it is interesting to note that internal casts of fossil shells can be found revealing the structure of the interior of the shells. Organic structures left behind by animals that have disappeared (*e.g.* worm borings) are called *trace fossils*.

It will be readily understood that animals with hard parts form fossils more readily than those without, the soft parts decaying more rapidly. Rapid burial is also a factor aiding preservation—thus the fossils of aquatic animals are more common than those of terrestrial ones. It should be generally noted also that for every living thing whose fossil has left a record of its existence there must have been many others which have not left a trace.

2. Early ideas about fossils. Fossils have been known and have excited man's interest from very early times and we know that the Greeks and Romans speculated about them. Many curious notions developed: for example, many believed fossils were the relics of creatures that were overwhelmed by the Biblical Flood; others thought that fossils were plants and animals in the making; yet others seem to have regarded fossils as "Nature Jokes!"

One of the first to take fossils seriously and to accept them as the actual remains of organisms which had once had living existence was the Renaissance artist-scientist Leonardo da Vinci. Examining the rocks of the Alps, he came to the conclusion that the strata and their contents must have once lain underneath the sea. He was forced to the conclusion that the fossil forms found in the rocks were the preserved remains of former living organisms which, in one way or another, had become entombed in the rocks.

Later, in the seventeenth century, Steno (actually Stensen, a Dane resident in Italy) made a fine collection of fossils and began to develop ideas of life sequences from the study of these and the strata in which they occurred. Yet it was not until the eighteenth century that they came to be accepted for what in truth they are, and indeed it was not until the latter half of the nineteenth century that people eventually accepted that they represented a record of the gradual evolution of life on earth.

OCCURRENCE AND CLASSIFICATION

3. The occurrence of fossils. *Fossils do not occur commonly in all rocks.* Igneous rocks, which were originally in a molten state and came from beneath the crust or the base of the crust, by their very nature seldom contain fossils although reference has been made above to the existence of moulds in some volcanic rocks. Again metamorphic rocks have been so altered by heat and pressure that any fossils that may have been in them are usually destroyed. *The only rocks which commonly contain fossils are sedimentary rocks.* It should be noted, however, that many stratified rocks have no fossils at all; this may be either because of the special conditions under which they were laid down or because the life forms were not suitable for fossilisation.

The oldest rocks on earth contain very few traces of life. This may be because life was in its earliest stages of evolution and only a few simple forms had evolved; it may also be due to the fact that early life was probably essentially soft-bodied and so seldom fossilised; and finally in most cases very old rocks have been so deformed that former traces of fossil life may well have been obliterated.

4. The importance of fossils. The single most important thing about a fossil is that it was once a living organism or associated with a living organism. Because of this *it provides a link in the long chain of evolving life.* Thus fossils are of prime importance in the study of the *evolution of life* on the earth, but they are also important as *tools in the study of historical geology or stratigraphy.*

It is reasonable to assume that the particular fossils occurring in each layer or stratum of rocks represent the types of life flourishing at the time of its formation. It is also clear that since life is constantly evolving and changing the fossils found in rocks of different ages would reflect these changes. Thus a bed or group of beds of a particular age would be characterised by a particular assemblage of fossils. The importance of this was first realised by Georges Cuvier in France and William Smith (Father of English Geology) in England in the early part of the nineteenth century. Smith saw that if a particular group of fossils were found in a certain stratum, and the same fossils were also found elsewhere in another stratum of similar type, then the latter must be of the same geological age as the former, even though they may be located far apart. Using this as one of his principles, Smith produced the first ever geological map of England and Wales, in 1815. Thus by a study of fossils it became possible to *correlate* rocks of widely separated regions.

A third importance of fossils is that they may *indicate the mode of origin* or *conditions under which the strata are formed and laid down*—that is they are essential to the studies of *palaeoecology* and *palaeogeography*. For example, if a bed of rock contains remains of marine organisms it is evident that it was laid down in the sea. But it may also be possible from a study of the fossils to obtain indications of the water depth and even the water temperature. Chemical studies of shell compositions (oxygen isotope studies) are proving very useful in this latter respect. It has proved important to establish the relationships of organisms to their environment (palaeoecology) because variations in stratal faunas (*the biofacies*) have often proved due to varying environmental conditions rather than to their being of differing geological ages. Faunas unable to tolerate varying environmental conditions are often described as facies sensitive.

5. The classification of fossils. Whenever a large amount of scientific information is amassed it becomes necessary to classify it or arrange it in an orderly way so as to simplify the task of extracting or adding data. *The classification* commonly used for fossils is a biological one whereby organisms are arranged into groups based upon their physical resemblances and evolutionary relationships. The outlines of this classification were first established by the Swedish scientist, Carl von Linné (Latin, Linnaeus) in the eighteenth century at which time Latin was used for scientific description. Thus it is that we find many fossil names are derived from the Latin or Greek languages. The classification groups plants and animals into a number of units of decreasing importance, typical examples of which are shown below:

	Modern Man	*Common oak*
Kingdom	Animalia	Vegetabilia
Phylum	Chordata	Spermatophyta
Class	Mammalia	Dicotyledoneae
Order	Primatida	Fagales
Family	Hominidae	Fagaceae
Genus	Homo	Quercus
Species	sapiens	robur

PROGRESS TEST 5

1. What is a fossil? (1)
2. How did people interpret fossils in early times? (2)
3. In what ways is the study of fossils associated with da Vinci and Steno? (2)

4. "Fossils do not commonly occur in all rocks." Explain why. (3)

5. What is the importance of fossils to the geologist? (4)

6. Why is William Smith often called the "Father of English Geology"? (4)

7. Who is thought to be responsible for first recognising fossils for what they are? And who was the first to draw up a classification of fossils? (2, 5)

8. Explain the meaning of the following terms: trace fossil, palaeo-ecology, biofacies. (1, 4)

CHAPTER VI

FOSSIL GROUPS

ANIMAL PHYLA AND PLANT DIVISIONS

Out of the fourteen or so animal phyla that exist, the marine invertebrates (animals without backbone) have left the best fossil record and provide the most commonly found fossils. Accordingly a short description of these is given in this chapter. In addition an outline is provided of the evolutionary development of the vertebrates (backboned animals) and the plants. The groups are:

Phylum Arthropoda   Class Trilobita (Trilobites)
Phylum Chordata (or Coelenterata)   Class Graptolithina (Graptolites)
Phylum Coelenterata   Class Anthozoa (Corals)

Phylum Brachiopoda
  (Brachiopods)

Phylum Mollusca   { Class Cephalopoda (Nautiloids, Ammonoids, Belemnoids)
Class Lamellibranchia (Lamellibranchs)
Class Gastropoda (Gastropods)

Phylum Echinoderma (Echinoids and Crinoids)
Phylum Porifera (Sponges)

Phylum Chordata, Sub-Phylum Vertebrata   { Classes Agnatha
Placodermi
Chondrichthyes } Fish
Osteichthyes
Amphibia
Reptilia
Aves (Birds)
Mammalia

Plant Kingdom   { Divisions   Thallophyta (Algae, Fungi, etc.)
Bryophyta (Mosses, Liverworts)
Pteridophyta (spore-bearing plants)
Spermatophyta (seed-bearing plants)

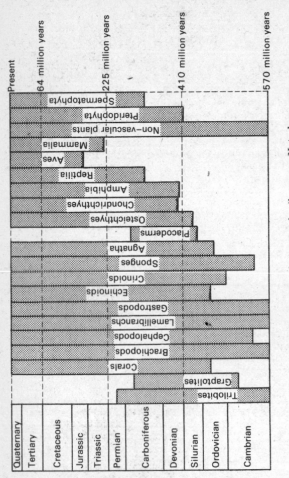

FIG. 8.—*Stratigraphical range of common fossil groups.* Note how some early groups died out, how some appeared relatively late in geological time, and others have persisted through prolonged geological ages right up to the present day.

ANIMALS

1. Trilobites. Belonging to the Phylum Arthropoda, the trilobites had a segmented body protected at the back by a *dorsal shield* or exoskeleton. It is this which is found fossilised. It is divided into a head region (cephalon), body (thorax), and tail (pygidium), the latter two also being divided by furrows into a number of segments. Characteristically the trilobite is also divided into three parts (tri-lobed) by two longitudinal furrows which separate an axial region from lateral (pleural) areas (*see* Fig. 9a).

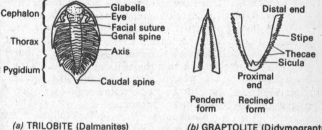

(a) TRILOBITE (Dalmanites)   *(b)* GRAPTOLITE (Didymograptus)

Fig. 9.—*Trilobites and graptolites.*

We cannot be absolutely certain of the mode of existence of trilobites but many of them must have lived on the sea floor crawling in the silts and muds rather like some modern crabs. The morphology of others, however, leads us to believe that they were adapted for swimming or for a floating planktonic existence.

The trilobites ranged from Cambrian until Permian times, but were at their most important during the Lower Palaeozoic where they have proved useful in stratigraphic correlation.

2. Graptolites. Graptolites are an extinct group of marine organisms whose biological affinities have not been entirely resolved. Most palaeontologists would either place them in the Phylum Coelenterata (relate them to corals and similar organisms) or associate them with Protochordates (primitive relations of the vertebrates). Their skeleton (*rhabdosome*) is a colonial structure consisting of one or more branches (*stipes*) supporting cup-like structures (*thecae*) in which were housed the zooids. In some genera they occur only on one side of the stipe (*uniserial*) and in others on both sides (*biserial*). There is great variation in the number and shape of the stipes as well as in thecal size and shape. At the base (*proximal end*) of the stipe is a conical structure termed the *sicula*. Growth by budding is believed to have started from this. A thread like feature (*nema*) is attached to the sicula in some graptolites. The graptolites are subdivided into two groups—the *Graptoloidea* and the *Dendroidea*. The latter are characterised by large numbers of branches or stipes, whereas the former mostly have eight or less. The graptolites were a group that achieved a very wide geographical distribution and it seems most likely that they had a largely floating planktonic type of existence. It has been suggested, however, that the zooids were capable of generating small currents which might at least enable them to station themselves in the photic zone (zone of maximum light penetration where food supply would be more abundant). Although they are most frequently found in black shales, it is probable that this is because the black mud environments with their lack of strong currents and bottom-dwelling organisms were the best suited for the preservation of the rather delicate graptolite rhabdosomes.

As a whole the graptolites range from the late Cambrian to the Carboniferous Period, although they are of greatest importance as zonal fossils in the Ordovician and lower Silurian. They are especially suited for stratigraphical purposes since not only did they achieve a wide geographical distribution but they also underwent rapid evolutionary changes. In general the changes demonstrate the following trends:

(*a*) A reduction in the number of stipes.

(*b*) A change in the declination of stipes, the trend usually being from pendent to scandent (*see* Fig. 9b).

(*c*) A change in thecal shape and spacing; normally a trend from simple to more elaborate shapes and towards greater spacing of the thecae along the stipe.

3. Corals. Corals belong to the Class Anthozoa of the Phylum Coelenterata. In their simplest form they consist of a conical or cylindrical cup (*theca*) the upper part (the *calyx*) of which contains the polyp. The latter secretes and lengthens its skeleton also producing partitions or plates within it. Particularly notable among these are a series of radial plates (*septa*) and near horizontal partitions (*tabulae*). Corals may be solitary or colonial in habit. A colony of corals is termed a *corallum*, each individual within it being a *corallite*. Two subclasses of the Anthozoans in particular provide common fossils. These are the *Tabulate* corals and the *Rugose* corals. The former consist predominantly of colonial masses of slender corallites characterised by well formed tabulae and a general absence of septa. The latter may be either solitary or colonial in habit with each corallite usually being divided by a complex of plates (*see* Fig. 10) of which the septa and a central column (*columella*) are particularly important. The order of secretion of these septa gives rise to *bilateral symmetry* rather than a perfect radial arrangement.

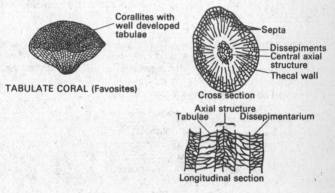

Corallites with well developed tabulae

TABULATE CORAL (Favosites)

Septa
Dissepiments
Central axial structure
Thecal wall

Cross section

Axial structure
Tabulae   Dissepimentarium

Longitudinal section

RUGOSE CORAL (Dibunophyllum)

Fig. 10.—*Corals.*

Modern corals are sessile marine organisms only some of which are reef formers. The reef corals are found in water less than 50 metres and above 25°C in temperature. Yet the non-reef corals may live in very different conditions and have been found in depths of over 2,800 metres and at temperatures as low as 4°C. It is believed that there are representatives of both reef corals and the deeper water corals among fossil forms. It should be emphasised, however, that many fossil reefs are unlike the present day ones in that they consist of swept up debris in which coral may be only one of several components. The name *bioherm* has been adopted as an all-embracing term for every kind of organic reef-like structure.

Corals range from the Ordovician up to present times. Tabulate corals are the oldest forms and are found almost throughout the Palaeozoic strata. The Rugosa appeared only very slightly later but by upper Palaeozoic times had become the dominant group. Yet the Rugosa and most of the Tabulata had become extinct by the opening of the Mesozoic era where new groups such as the Hexacorals and Octocorals made their appearance to carry on the line of evolution to the present day. In Britain corals became particularly well known as stratigraphical fossils in the Lower Carboniferous strata. A geologist called Vaughan in fact originally zoned the Lower Carboniferous sequence of South West England very largely on the basis of its coral faunas although subsequently this zoning has not proved entirely ideal.

4. Brachiopods. Commonly known as lamp shells, the brachiopods (*see* Fig. 11) are a phylum of shellfish whose structure consists of two unequally sized shells (*valves*) often hinged

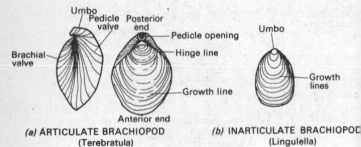

Fig. 11.—*Brachiopods.*

and held together by muscles. The larger valve (*pedicle valve*) commonly has a beak like projection (*umbo*) at the posterior end. This has a perforation (*pedicle opening*) through which a fleshy stem (*pedicle*) emerges to anchor the organism to the sea floor. The smaller valve (*brachial valve*) may have a looped or other type of skeletal structure (*brachidium*) on its interior side which in life served as a support for part of the organism. Brachiopod shells show a great deal of variation in shape and in shell ornament. Growth lines, ribs or costae, and spines are some common types of ornament. The phylum is divided into two classes: the *Inarticulata* and the *Articulata*. Some of the more important differences between these are summarised below:

Articulates	*Inarticulates*
Valves are hinged.	Valves are not hinged.
Calcareous shells.	Shells are mostly chitino-phosphatic.
Have some form of brachidium.	No brachidial structure.
Musculature is relatively simple.	Complex pattern of muscles to hold valves closed.

Brachiopods are marine organisms. After the larval stage most live fixed to the sea floor by means of the pedicle, although a few cement themselves to rocks and others may utilise their spines as a means of anchorage. Modern brachiopods are found in a great variety of water depths and water temperatures and it is probable that fossil forms similarly reflect a wide variety of conditions although perhaps the greater proportion of them were shallow water organisms.

The phylum ranges in time from the Cambrian to the Recent, although it was more strongly represented during the Palaeozoic and Mesozoic eras than at present. The Inarticulates dominate among the very earliest brachiopod faunas but they were very soon overtaken by the Articulates. Brachiopod fossil assemblages have been of use in the correlation of strata of several geological periods but like the corals their use is sometimes limited in that they may be facies sensitive (*see* v 4).

5. Molluscs. Molluscs are a very large phylum of organisms displaying such variation of shell form and living habit that superficially there often appears little similarity between its members. The phylum is subdivided into five classes, three of which—the *Cephalopoda*, the *Lamellibranchia* (Pelecypoda), and the *Gastropoda*—provide its most commonly found fossils.

(*a*) *Class Cephalopoda.* Most of the cephalopods which include the *Nautiloids* and *Ammonoids* (*see* Fig. 12(*a*)) have a shell (*conch*) which consists of a tube coiled into a flat spiral although a few may be straight or gently curved. Internally it is divided into a series of chambers (*camerae*) by transverse partitions (*septa*). In life the septa are pierced by a thin tube (*siphuncle*) around which they extend into protective necks (*septal necks*). In addition, where the septa join the outer shell, lines (*suture lines*) are produced, the patterns of which have proved very important in classification and evolutionary studies. Shells which are so tightly coiled that an outer coil or whorl envelopes inner ones are termed *involute* in shape whilst those showing no overlap are termed *evolute*. Involute shells have a well formed central depression (*umbilicus*). Finally cephalopods display a variety of shell ornament. This includes curved ribs of varying pattern, spiral or radial striae, tubercles, spines, ventral keel, and ventral groove or sulcus.

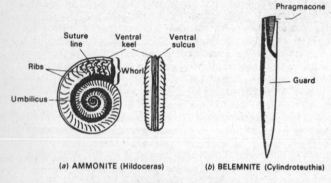

(*a*) AMMONITE (Hildoceras)   (*b*) BELEMNITE (Cylindroteuthis)

FIG. 12.—*Ammonite and belemnite.*

One common fossil cephalopod group not belonging to the nautiloids or the ammonoids are the *Belemnites* (*see* Fig. 12(*b*)). They have a solid calcareous cigar-shaped shell (*guard or rostrum*) at the top end of which is a conical cavity (*alveolus*) that contains a chambered tube (*phragmacone*) only occasionally found fossilised.

The modern representatives of the cephalopods are the squids, octopods, and most importantly, because of its resemblance to fossil forms, the genus Nautilus. By analogy with these modern types it seems certain that the cephalopods were free floating or swimming marine organisms. Nautilus has its chambers filled with gas which apparently can be varied to produce the required buoyancy for rising or sinking, and it is seldom found in water deeper than 700 metres. It is generally thought that the fossil forms were similarly shallow-water dwellers probably being found in continental shelf areas. The abundance of some cephalopods in certain rocks suggests that like modern squids they may also have been gregarious in habit.

Nautiloids are the longest-ranging members of the cephalopoda having existed from the Cambrian to the Recent. The ammonoids evolved from them in the early Devonian, underwent rapid expansion and evolution throughout the Mesozoic era, and became extinct at the end of the Cretaceous Period. Because of this rapid evolutionary change and their wide geographical dispersal the ammonoids have proved excellent fossils for correlation purposes; they have been widely used in this country for zoning Upper Carboniferous, Jurassic and Cretaceous rocks. The belemnites are much less important. They developed in the late Palaeozoic and like the ammonites flourished during the Mesozoic only to become extinct at the beginning of the Cainozoic.

(b) *Class Lamellibranchia*. Lamellibranchs (*see* Fig. 13(*a*)) are bivalved animals, most of which have equally sized left

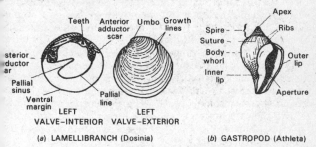

FIG. 13.—*Lamellibranch and gastropod.*

and right valves. The valves are hinged together by interlocking *teeth* and *sockets* of varying patterns. They were opened and closed by ligaments and muscles, the scars of which remain to show their position of attachment to the shell interior. There are normally two well marked *adductor scars* on the interior of each valve joined together by a line (*pallial line*) which marked the edge of the fleshy mantle of the organism. Some lamellibranchs have only one muscle scar and are said to be *monomyarian*. Those displaying an indentation in the pallial line (*pallial sinus*) are termed *sinupalliate*. Externally the valves are ornamented by *growth lines* and sometimes also by *ribs*. The ribs radiate out from a swollen beak-like feature on the dorsal margin of the valve (*umbo*). Except for a few genera the umbo is directed towards the anterior end of the valve. Other useful guides to shell orientation are given by the muscle scars and pallial sinus. Where the adductor scars are unequal in size the larger one is on the posterior side, and if there is only one scar then this is posterior in position. Similarly the pallial sinus, if present, is on the posterior side of the valve.

Lamellibranchs are aquatic and are found in fresh, brackish and marine waters. They are mostly bottom dwellers but are noted for the way in which they have adapted to a variety of environments and living habits. Some are *active* crawling or partially swimming over the sea floor. Others are *burrowing* forms, boring into rock and wood as well as the sea bottom. Yet a third group are *sedentary*, anchoring themselves by means of byssal threads or by cementing one valve to the floor. It is these latter cemented forms (*e.g.* oysters) that develop in an inequivalved fashion. Whilst most lamellibranchs live in the littoral zone, they have also been found at very great depths.

The stratigraphical range of the lamellibranchs is from the Ordovician to the Recent. Generally their preservation in the Palaeozoic is poor although they have been used for the correlation of the Coal Measures. Many new forms evolved during the Mesozoic to replace the Palaeozoic ones and although their expansion has slowed during the Cainozoic they are more abundant and better preserved in the strata of this era than the preceding two.

(c) *Class Gastropoda*. Gastropods (Fig. 13(*b*)) are the most numerous molluscs and are found in the widest variety of environments. They have a single shell or conch which in most cases is coiled into a *helicoid spiral*. Each turn of the shell is termed a *whorl*. Where the whorls are coiled in a clockwise fashion and the shell opening (*aperture*) appears on the right-hand side, the coiling is said to be *dextral*, and when anti-clockwise and on the left-hand side, *sinistral*. The whorl containing the aperture is the one in which the organism partially lives and is consequently termed the *body whorl*. The other whorls form the *spire*, the point of which is the *apex*. In some genera, the inner parts of the whorls coalesce to form an axial pillar known as a *columella*, whilst in others a central space (*umbilicus*) is left running axially inside the whorls. The shape of the aperture is often helpful in identification. Sometimes the margin of the aperture is characteristically notched and it may even be extended out into a narrow tube-like form (*siphon*). Finally the external ornament is also very varied consisting of forms such as ribs, striae, knobs, and spines.

Present-day gastropods are found in many environments, being marine, fresh water, and land dwellers. Most of the aquatic forms live in shallow well-lit waters where they are essentially free-crawling bottom dwellers. It is often not possible to demonstrate the exact environmental conditions reflected by the fossil forms although in many cases it is possible to show whether they were fresh water or marine forms.

In spite of their great stratigraphical range—from the Cambrian to the Recent—gastropods have not proved very useful fossils. Palaeozoic specimens are often not well preserved and it is in the Cainozoic, where they are at their most numerous, that gastropods have been of most use.

6. Echinoderms. The Phylum Echinoderma contains two particularly common classes of fossil—the *Echinoids* and *Crinoids*.

(*a*) *An echinoid* has its body encased in a firm globular shell

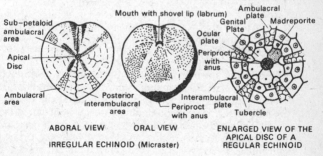

ABORAL VIEW   ORAL VIEW   ENLARGED VIEW OF THE
APICAL DISC OF A
REGULAR ECHINOID

IRREGULAR ECHINOID (Micraster)

Fig. 14.—*Echinoids*.

(*test*) which is composed of numerous calcareous plates. At the centre of the upper or *aboral* surface is a complex of small plates known as the *apical disc* (*see* Fig. 14). Radiating out from this are columns of interlocking plates. They are arranged in ten distinct zones, five of which are termed *ambulacral areas* and five *interambulacral areas*. The arrangement of these areas produces the pentameral symmetry so characteristic of all echinoderms. The ambulacral plates are normally smaller than those of the interambulacral areas. They also contain small holes (*pore pairs*) through which the organism's tube feet (*podia*) are passed to the exterior. The interambulacral plates are not perforated in this way but they commonly contain spine bases (*tubercles*). The *spines* which are fixed to these usually become detached in fossil forms. On the under side or *oral surface* of the test is the *peristome*. This is a small roughly circular structure of numerous tiny plates at the centre of which is the *mouth*. The peristome plates are also usually lost during fossilisation. The echinoids may be divided into *Regular* and *Irregular* groups. Some of the differences between these are summarised below.

Regular	Irregular
Periproct with anus in apical disc.	Anus outside the apical disc—usually in the posterior interambulacral area.
Peristome at the centre of the oral surface.	Mouth may be central but is often in an anterior position.
Test is roughly circular in shape with good radial symmetry.	Test is only bilaterally symmetrical.

The echinoids are entirely marine, living as free-moving scavengers on the ocean floors. Some forms burrow into the bottom sediments and a few even cut hollows in rocks to protect themselves from damaging wave action. Most echinoids live in waters less than 300 metres deep, but they have been found at much greater depths. They are typically gregarious in habit.

The geological range of the echinoids is from the Ordovician to the Recent. They are rather rare in Palaeozoic rocks, but expand rapidly in numbers during the Mesozoic. Indeed, it is during the Jurassic period that the irregular forms first make their appearance. Echinoids remain a flourishing group today. Although many fossil echinoids have a limited vertical distribution, their restricted geographical distribution has often spoiled their suitability for correlation work. In Britain they are best known for their use in the correlation of the Chalk.

(*b*) *Crinoids* (*see* Fig. 15(*a*)) are commonly known as sea-lilies although they are animals and not plants. They consist of three distinct parts—the *roots*, *stem* and *crown*. The roots and stem are composed of a number of small, articulated plates called *ossicles* or *columnals*. Their most common shape is a circular or pentagonal disc. At the top of the stem is the

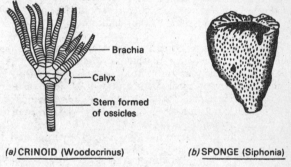

(a) CRINOID (Woodocrinus)   *(b)* SPONGE (Siphonia)

Fig. 15.—*Crinoid and sponge.*

crown, which is divided into a cup (*calyx*) and five arms (*brachia*). During life the arms were utilised in procuring food and were spread out from the margins of the calyx. When fossilised they often appear tightly closed.

Crinoids differ from the echinoids in that they mostly live fixed to the sea floor. They commonly live in clusters in shallow water and in particular around reefs. Thus some limestones forming in such environments are almost entirely composed of crinoid fragments (*crinoidal limestones*).

Crinoids range from the Ordovician to the Recent. They are the most abundant echinoderms of the Palaeozoic era, but become less important during the Mesozoic. At the present day there are still about one hundred genera surviving. Crinoids are poor stratigraphical fossils because of their limited geographical distribution, but Marsupites, one of the uncommon free-floating varieties, has been used in the correlation of the Chalk.

7. Sponges. Sponges (*see* Fig. 15(*b*)) form the Phylum Porifera. They are fairly simple, multicellular, aquatic organisms. Living attached to the floor, they secrete a skeleton of silica or calcium carbonate. It consists of very small intermeshed rods or *spicules* which frequently fall apart when the organism dies. However in some cases the structure may remain intact. In many instances fossil sponges are plant-like in appearance with a roughly cylindrical or conical shape. Although they range from the Cambrian to the Recent the sponges are not important stratigraphical fossils. In Britain they are particularly common in some parts of the Chalk.

8. Vertebrates. Vertebrates are backboned animals belonging to the Phylum Chordata. Although some of their aquatic representatives yield reasonable numbers of fossils, on the whole specimens are much rarer than those of their invertebrate counterparts. Nevertheless the vertebrates embrace some of the most successful of all animals and thus a very brief review of their evolutionary development is given below.

The first definite vertebrates were fish. They appeared during the Ordovician but did not become widespread until the late Silurian and Devonian. The earliest of these, the *Ostracoderms*, lacked proper jaws and had no paired fins. They were also covered with an armour of bony plates. The *Placoderms* which appeared in the late Silurian showed some advances. They had developed primitive jaw structures and paired fins but many of them still remained heavily armoured in the head region. The most successful group of fish were the *Osteichthyes* or bony fish. They evolved in the Devonian and remain a flourishing class today. They are characterised by having internal bony skeletons and scale covered bodies. A few also have lungs. There is also one other class of fish still in existence—the *Chondrichthyes*. Ranging from the Devonian, they have skeletons of cartilage rather than bone and include forms such as sharks and rays.

The *Amphibians* were the first class of the vertebrates to

invade the land. It is thought that they evolved from some of the bony fish towards the end of the Devonian period. At that time the fish had developed a bone support in their fins and the ability to breathe air through internal nostrils. This may have enabled them to flounder over muds from one stretch of water to another. In this way perhaps, the amphibian mode of life gradually emerged. Their real spread took place during the Carboniferous. Limbs replace fins, the backbone became much stronger, and adult amphibians developed proper lungs.

The domination of the amphibians on land did not survive long, however, for as early as the end of the Carboniferous came the appearance of the *Reptiles*. They were clearly better suited to the terrestrial environment than amphibians. Of particular importance in this connection was the fact that their eggs, laid on land, were covered by a tough protective skin and contained a supply of food. The early reptiles were not more than two metres long and were of a clumsy build. Perhaps one of the more notable changes was in the skull. The flat skull of the amphibians was replaced by a rather higher and narrower skull.

The reptiles expanded and diversified in a remarkable way during the Mesozoic and adapted to many environments. Some like the *Ichthyosaurs* and *Plesiosaurs* developed stream-lined fish-like bodies and clearly were marine carnivores of a very large size. Some of the *Dinosaurs* or *Archosaurs* are also well known because of their great size and extraordinary appearance. These terrestrial forms included both carnivores and herbivores. The former had a bipedal gait with their hind limbs being much more massive than the front ones. They also tended to develop large skulls with powerful jaws and teeth. Among the better known of these are *Tyrannosaurus*, which was about fifteen metres long and six metres high, and *Allosaurus* which was only slightly smaller. Some of the herbivores were even larger, *Brontosaurus*, *Diplodocus*, and *Brachiosaurus* were all over twenty metres long and the latter probably weighed around fifty tons. Herbivores such as these were characterised by a quadruped gait, very long necks and tails, and usually by relatively tiny skulls. Not all dinosaurs of course were so large, many in fact not being much larger than a dog. Some reptiles apparently developed the ability to fly or perhaps glide through the air. The *Pterodactyls* and *Rhamphorhynchoids* for example had wing-like membranes attached to greatly length-ened forelimbs.

Many of the reptiles including the dinosaurs became extinct at the end of the Mesozoic. During the Cainozoic their habitats were quickly invaded by the fast evolving *Aves* (birds) and *Mammals*. Yet the origins of both of these groups were in the Mesozoic. The earliest birds for example were found in the Jurassic. At that stage their appearance was very similar to small flying dinosaurs. However, the remains of *Archaeopteryx* and *Archaeornis* found in Germany revealed impressions of feathers and so are usually regarded as ancestral birds. The earliest mammals occur at the beginning of the Triassic period. At that time there were a number of mammal-like reptiles in existence and it is difficult to determine precisely when the mammalian characteristics of warm-bloodedness and viviparous reproduction arose. Like the reptiles before them, the mammals have adapted to a very wide range of environments and have become the dominant class of the Cainozoic era.

It was at the beginning of this era that the mammalian group of the Primates evolved from tiny tree dwelling insect eaters to lead on eventually to the emergence of man. The first undoubted hominids were the *Australopithecines*, a race of ape-like men living in Africa between five million and one million years ago. Yet there is considerable doubt as to whether these were the direct ancestors of modern man for other controversial fossil discoveries in Africa have suggested that there were other hominid varieties, perhaps less ape-like, in existence at the same time. The earliest fossils which appear to be the definite forerunners of *Homo sapiens* date from about 800,000 to 500,000 years ago. They are the remains of *Homo erectus* (includes Java man, Pekin man, etc.) which have been located in Asia, Africa and Europe. In fact it was not until the last 100,000 years that men similar in type to present-day man gradually emerged.

PLANTS

9. Multicellular plants. Multicellular plants can be divided into two categories—the *non-vascular* and *vascular*. The former have little cell differentiation and are largely aquatic. The latter develop cell specialisation enabling food to be transmitted from one part to another and are mainly terrestrial.

10. Non-vascular plants. The most common fossils of the non-vascular group are *algae* (seaweeds). They are the oldest known fossils, some occurring in Pre-Cambrian rocks over 2,500 million years old. The *calcareous algae* secrete calcium carbonate and develop as a fine mat-like covering over reef debris. They thus act as a binding agency and their remains are common in bioherms.

11. Vascular plants first evolved at the end of the Silurian period and the beginning of the Devonian. Initially there was no proper differentiation into root and leaf systems, but by Carboniferous times a wide variety of spore-producing (*pteridophytes*) and seed-bearing ((*spermatophytes*) plants were colonising the land.

At this time the present-day areas of Europe and North America were drifting through the tropics. Hot, low lying swamps abounded upon which forests of evergreen trees and ferns flourished (*see* Fig. 16). The accumulation of plant debris in such environments ultimately led to the formation of the Carboniferous coal seams. These plants continued their

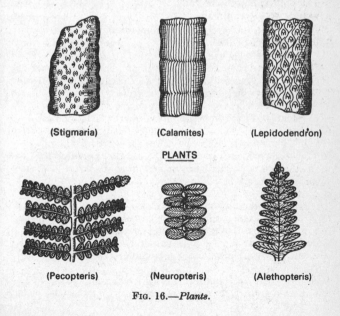

(Stigmaria) (Calamites) (Lepidodendron)

PLANTS

(Pecopteris) (Neuropteris) (Alethopteris)

Fig. 16.—*Plants.*

spread during the Mesozoic era but by the Cretaceous the flowering plants (*angiosperms*) had appeared. The latter quickly developed to dominate vegetation during the Cainozoic. In particular there was a rapid spread of many grasses as drier climates developed. These formed a new source of food supply and probably affected the course of Tertiary mammalian evolution.

MICROFOSSILS

The fossils described above can usually be detected reasonably easily with the naked eye. They are sometimes referred to as *macrofossils* on account of their size. In contrast there are a number of fossils whose skeletons are so small that they can often only be recognised with the aid of a microscope. These are referred to collectively as *microfossils*. They have proved to be of the utmost importance in applied geology. Much of the data obtained in the search for minerals comes from borehole cores and chippings. Macrofossils are only occasionally found in such limited samples but microfossils are much more common. Moreover the microfossils have often proved to be equally reliable or even more reliable than macrofossils for correlation. Representatives of both animal and plant kingdoms are found among microfossils.

Perhaps the best known from the animal kingdom are the *Foraminifera*, unicellular organisms belonging to the Phylum Protozoa. They have minute shells or *tests* of various shapes which are divided into numerous *chambers*. Fossil forms either have calcareous tests or arenaceous tests, the latter consisting of small sand grains cemented together. Their time span is from the Ordovician to the Recent. *Radiolaria* are also members of the Protozoa which form abundant fossils. They have rather intricately shaped spiny silica tests that form a large proportion of some deep sea oozes. They range from the Cambrian to the Recent.

Their equivalents from the plant kingdom are the *Diatoms*. They are microscopic unicellular algae that exist in both fresh and marine waters. They also produce siliceous skeletons that contribute greatly to some deposits (*e.g.* diatomaceous earth). Stratigraphically they range from the Cretaceous to the Recent.

Finally one other collection of microscopic plant life which has received much attention in recent years is that comprising spores and pollen. The study of these has developed into a science termed *Palynology* which is proving of immense value in the reconstruction of Cainozoic environmental conditions. Although its use in pre-Cretaceous rocks is rather limited, spore studies have made valuable contributions to British Coal Measure geology.

PROGRESS TEST 6

1. With the aid of labelled diagrams describe a typical trilobite. (1)

2. In what strata are graptolites found? Why are they such useful stratigraphical fossils? (2)

3. Describe the differences between rugose and tabulate corals. Illustrate your answer with labelled diagrams. (3)

4. Outline the main differences between a typical brachiopod and a typical lamellibranch. (4, 5)

5. Give an account of the variation of morphology seen in the cephalopods. (5)

6. Discuss the adaptations to varying environments and modes of life shown by the lamellibranchs and the gastropods. (5)

7. Compare a regular with an irregular echinoid. (6)

8. Describe the morphology of a crinoid. Under what conditions did they live and what sorts of rock are they found in? (6)

9. Describe the general sequence of evolution of either the plants or the vertebrates. (8, 9)

10. Give a short account of the various types of microfossil. (*Microfossils*)

CHAPTER VII

THE EARTH: ITS ORIGIN, STRUCTURE AND COMPOSITION

THE EARTH AND THE UNIVERSE

1. The universe. The study of the evolution of ideas regarding the origin of the earth and its relation to the solar system and of the latter to the universe is known as *cosmology*. From the very earliest times man was fascinated by the earth and the heavenly bodies, and pondered upon their origin. All the early civilisations developed their own particular versions of the origin of the earth and the nature of the universe.

Until the invention of the telescope man's ability to peer into space and explore the universe was very limited. During the past two or three hundred years the development of more powerful telescopes, the invention of the radio-telescope and developments in mathematics and physics have enabled man to explore space much more adequately and to theorise upon the nature and the origin of the universe.

This subject really lies outside the scope of geology, but it will be useful to summarise very briefly the main ideas about the universe held at the present time. There are two rival theories on the origin of the universe:

(a) *The superdense or "big bang" theory* advanced by Ryle propounds that the universe was created by a vast explosion of superdense matter; this produced the galaxies of stars which are scattered throughout space and which are flying apart from each other at the astonishing speed of 370 million m.p.h. (approx. 600 million km/h).

(b) *The steady state or continuous creation theory*, elaborated by Hoyle, Bondi and Gold, which rejects this explosive and finite beginning and puts forward the idea that matter is being continuously created; according to this theory new galaxies are born and compensate for those which are receding beyond man's ken.

2. The origin of the solar system. The various planets (and their satellites) which revolve around the sun form the solar system. The sun itself is but one of the 100,000 million stars belonging to the Milky Way Galaxy.

In ancient and medieval times it was thought the earth was the centre of the universe: this idea is known as the *geocentric theory*. Copernicus demolished this traditional belief and put the earth in its proper place as merely one of a number of planets in a sun-centred system: this concept is known as the *heliocentric theory*, although it is now no longer a theory but an established fact.

During the past two hundred years several theories concerning the earth's origin have been advanced:

(*a*) The *nebular theory* of Kant and Laplace in the eighteenth century postulated the origin of the solar system from a hot, rotating, gaseous mass which, contracting as it cooled, threw off rings of matter which condensed to form planets.

(*b*) The *planetesimal hypothesis* of the Americans, Chamberlin and Moulton, put forward at the beginning of this century, supposed that the planets had come into being through the coalescence, by gravitational attraction, of planetesimals or fragments of solar material.

(*c*) The *tidal theory* of Jeans assumed that a wandering star approached the sun thereby producing tidal distortion in the sun and resulting in a cigar-shaped filament of solar matter being pulled away from it, which in due course materialised into concentrations of planets.

(*d*) The *twin-sun theory*, explained by Hoyle, which supposes the solar system to have been derived from the sun's former twin which exploded and shattered, some of the material subsequently becoming concentrated into planets by gravitational attraction.

It should be noted that there are many theories in the field but that none of them is entirely satisfactory or meets all the objections which are put forward. A modified nebular theory is rather in vogue at the present time.

THE EARTH'S SHAPE AND SIZE

3. The shape of the earth. To early man, with his limited powers of perception, the earth was regarded as being flat: in other words, it had a plane surface. This idea of a disc-like earth was common to all the peoples of the ancient world until the Greek philosophers began to teach otherwise. The idea of a spherical earth, however, was rejected by most scholars during medieval times and the old idea of a flat earth persisted until the sixteenth century when Copernicus' heliocentric theory and Da Gama's circumnavigation of the globe caused men to think afresh about the form of the earth.

Newton, arguing that the earth was a rotating body and as such would be subject to centrifugal forces and therefore tend to bulge out in those parts furthest removed from the axis of rotation, *i.e.* in its equatorial regions, deduced that the earth was roughly orange-shaped: in other words, it departed from perfect sphericity through slight polar flattening. Later measurements confirmed Newton's suggestion that the earth was compressed upon its polar axis and bulged slightly at the equator. Such a form with its two deviations from the perfectly spherical form is termed an *oblate spheroid* or an *ellipsoid of revolution*.

In 1903, Sir James Jeans calculated that the earth was more nearly pear-shaped than orange-shaped and the latest measurements confirm that Jeans was correct. To surmount the difficulties of describing precisely this slightly irregular spherical form of the earth the term *geoid* was invented, although this does not get us very far since it simply means "earth-shaped"!

4. The size of the earth. Geodesy is the science which studies, determines and measures the exact shape and dimensions of the earth. The latest geodetic survey together with satellite observations gives the earth's dimensions as follows:

equatorial diameter	—	7,926·42 miles (12,755 km)
polar diameter	—	7,899·83 miles (12,713 km)
equatorial circumference	—	24,902·44 miles (40,075 km)
meridional circumference	—	24,860·53 miles (40,008 km)

This departure from the perfect spherical form is only very slight and it must not be over-emphasised: indeed, viewed from space, say from the moon, the earth would appear as a true sphere.

5. Surface irregularity. In spite of its spherical form, it is obvious that the outer surface of the earth is by no means smooth. There is a pronounced irregularity of relief produced by high mountains and ocean deeps. The land surface culminates in Mt Everest which is 29,028 feet (8,848 m) high while the ocean floor descends to its greatest known depth in the Mariana Trench, off the island of Guam in the Pacific, which lies 37,800 feet (11,521 m) below sea level. The total vertical range of the surface of the earth is thus 66,941 feet (20,404 m), or just over 12 miles (19·3 km). This surface irregularity, though impressive to the people living on the earth's surface and of great human importance, is only slight when compared with the earth's dimensions as a whole: the maximum vertical range of 12 miles (19·3 km) is only about 0·154 per cent of the earth's diameter.

Figure 17, known as a *hypsographic curve*, shows the areas of the earth's crust between successive levels from the highest

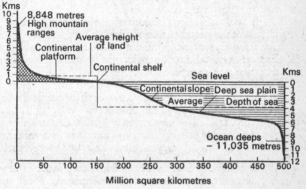

After A. Holmes

Fig. 17.—*Hypsographic curve.* This shows the areas of the earth's crust between successive levels from the highest mountain peak to the greatest known ocean deep.

mountain peak to the greatest known ocean deep. Compare the average height of the land surface with the average depth of the sea; note, also, the proportion of land surface to ocean surface—29·2 per cent and 70·8 per cent respectively.

THE EARTH'S COMPOSITION AND STRUCTURE

6. What lies beneath the crust? While man is able to see upwards into space for millions of miles, the deepest penetrations downward (by boring) are only in the region of about 14,000 feet (4,270 m). Thus, so far, man has really only scratched the surface of his planet and he has only penetrated to a minute depth—of less than 0·001 per cent of the distance from the surface to the centre of the earth.

In order to improve our knowledge of the sub-crust, American scientists planned to drill a hole, some six miles deep, through the crust to the material below. The project, the Mohole Project, was eventually abandoned after costly early work. Subsequently, however, a programme of deep sea drilling to obtain cores of ocean crust from many geographical locations was instituted (the J.O.I.D.E.S. project), and has been carried out by the ship *Glomar Challenger*. Valuable information so obtained has done much to verify our ideas about the behaviour of the earth's crust and the rocks beneath.

Although we do not know for sure what lies beneath and what the condition of the earth's interior is like, intelligent interpretation, supported by various kinds of evidence, has enabled the geologist to theorise about the inside of the earth: its composition, structure and nature.

7. Composition and structure. The earth is thought to be composed of a series of onion-like layers or "shells" of different materials (Fig. 18). Authorities vary slightly in their views about the number of layers, their constitution and thickness, but the following is a fairly generally accepted interpretation.

(*a*) *The crust*, or outer layer, which is composed of crystalline rocks with sedimentary rock cappings. This outer layer is about 20 to 30 miles (32·2 to 48·3 km) deep and the rocks have a specific gravity of between 2·0 and 3·0. This outer zone is separated from the zone beneath by a distinct break or discontinuity known as the *Mohorovicic discontinuity*. It is usual to distinguish two layers in this crustal zone:

(*i*) the outer layer, made up of sedimentary and granitic rocks, known as *sial* (from *Si*, silica and *Al*, alumina); and
(*ii*) the under layer, made up of dense basaltic rocks, known as *sima* (from *Si*, silica and *Ma*, magnesia).

(*b*) *The mantle*, or middle layer, composed of ultrabasic rocks. It has sometimes been suggested that the mantle consists of the mineral olivine, which is a heavy silicate of iron and magnesium. The rocks of the mantle have a specific gravity of between about 3·0 and 5·0. The mantle, which extends downwards to a depth of about 1,800 miles (2,898 km), is separated from the centre or core of the earth by another discontinuity, known as the *Gutenberg discontinuity*. Sometimes the mantle is sub-divided into an inner and an outer mantle.

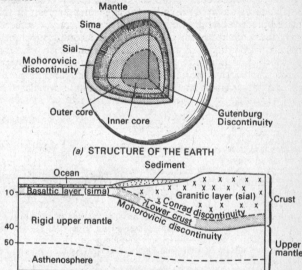

(a) STRUCTURE OF THE EARTH

(b) SCHEMATIC SECTION OF THE LITHOSPHERE

FIG. 18.—*(a) A "cut-out" of the earth illustrating the internal structure, of onion-like layers, of the earth. (b) A schematic section of the lithosphere.*

    (c) The core of the earth, known as the *centrosphere* or *barysphere* (*i.e.* heavy sphere) lies below 1,800 miles (2,898 km). It is thought probably to be metallic and to consist of iron and nickel: hence the core is sometimes referred to as *nife* (*Ni*, nickel and *Fe*, iron). The specific gravity of the core ranges from about 5·0 to 13·0. Recent research suggests the barysphere may consist of two layers:

    (*i*) an outer core which is liquid or in a plastic state; and
    (*ii*) an inner core which is most likely solid.

8. The nature of the earth's interior. The earth's outer crust is solid but what is the nature of the earth beneath the crust and at its centre? We can really only surmise what it is like. However, we do know beyond any shadow of doubt that both temperature and pressure increase with depth as one penetrates below the earth's surface. At the earth's centre the temperature must be very high indeed and the pressure very great. Under the high temperatures prevailing at the centre we can imagine that the material would be in a liquid state or perhaps gaseous but, because of the terrific pressure, it is in a condition which we cannot very readily comprehend. The geophysicist, whose job it is to explore and interpret the earth's interior, asks us to imagine the earth as a ball which is as strong and as rigid as steel and yet, at the same time, possesses a quality of plasticity!

PROGRESS TEST 7

1. Describe briefly the two principal rival theories relating to the origin of the universe. (1)

2. Explain briefly what is meant by (a) the geocentric theory and (b) the heliocentric theory of the solar system. (2)

3. Name, and briefly describe, four theories concerning the origin of the solar system. (2)

4. With what cosmological ideas are the names of the following scientists connected—Hoyle, Copernicus, Jeans, Ryle? (1, 2)

5. Write a brief account of the size and shape of the earth. (3, 4)

6. Write an account of the composition, structure and nature of the earth. (6, 7, 8)

7. Explain the meaning of the following terms: geoid, planetesimals, barysphere, sial, cosmology. (1, 2, 3, 7)

8. Write short explanatory accounts of the following: nebular theory, hypsographic curve, Mohorovicic discontinuity. (2, 5, 7).

9. Draw an annotated diagram to show the internal structure of the earth. (7)

10. Define the fields of investigation of (a) the cosmologist and (b) the geophysicist. (1, 8)

CHAPTER VIII

SIMPLE STRUCTURES, EARTH-BUILDING FORCES

1. The forces in landscape formation. The features of the earth's surface, whether continents or oceans, hills or valleys are not stable and changeless; slowly but surely they undergo alteration. Changes are, in fact, continuously occurring both beneath and on the surface of the earth. Usually these changes are so slow as to be imperceptible but occasionally, as when volcanoes erupt, earthquakes happen or landslides take place, we can witness sudden changes taking place.

The surface relief or topography, which is the outward expression of structure, results from the action of two forces:

(a) *Tectonic forces, i.e.* those giving rise to land uplift or subsidence, folding, fracturing and volcanic eruption; these forces may be said, in general, to be responsible for the major structural units of the earth's surface, *e.g.* mountain ranges, plateau blocks and the great plains.

(b) *Denudational forces, i.e.* those giving rise to the destruction, carving, moulding and smoothing of the major relief features: they produce the intricate details of the surface topography, *e.g.* the hills, valleys, spurs, potholes, caves, sea-stacks, etc.

2. Orders of relief. At this point we might conveniently refer to what the geomorphologist terms the "orders of relief". The orders represent scales of decreasing magnitude:

(a) Relief features of the *first order* are the largest earth features—that is the continental areas and the ocean basins.

(b) Relief features of the *second order* may be thought of as the major structural units, *e.g.* mountain systems, plateau blocks.

(c) Relief features of the *third order* are the small, individual topographic features, *e.g.* isolated hills, sand-dunes and river terraces.

3. Diastrophism. The major relief features or structural units, such as mountain systems and plateau masses, have been mostly created through the action of internal forces operating beneath the crust. The term *diastrophism* is given to all such forces which disturb and dislocate the crust. Diastrophic forces produce two kinds of movement:

(a) *Epeirogenic movements:* these are vertical movements or continent-building movements. They result in the uplifting or depressing of large segments of the crust, *e.g.* the raising up of the African Plateau, the large-scale down-warping of the Aral-Caspian Depression.

(b) *Orogenic movements:* these are horizontal movements tending either to compress or extend the crust, compression involving folding and faulting, tension involving fracturing and faulting; such movements lead to the formation of mountains, *e.g.* the young fold mountains of the Alps and Andes and the massive block faulting of the East Australian Highlands.

4. Isostasy. Another force which the geologist must take into account is *isostasy*. The theory of isostasy is, simply put, that a state of equilibrium of balance exists in the earth's crust. It implies the idea that the continental land masses consisting of light material (sial) floating as it were, in a substratum of denser material (sima). If the continents are lightened in any way as, for example, if they lose some of their material through erosion or through the removal of an ice sheet, then the continental block will tend to rise or float higher in the sima.

This concept of isostasy is not easy to understand but can perhaps best be grasped by thinking of several blocks of wood of the same cross-section but of differing heights, floating in a tank of water. Each of the different blocks raises itself above the water level by an amount which is proportional to its length. A series of such blocks are said to be in a condition of *hydrostatic balance*. Now, if an inch, for example, was sawn off one of the blocks, that particular block would adjust itself in the water and a part of the block previously submerged would lift itself above the water level, while the block as a whole would not sink as deeply into the water as it did originally. In like manner, the denudation (or loss of weight by other means) of a landmass will upset the equilibrium of the land mass and cause it to rise up until the correct adjustment has been made. Conversely, of course, any additional weight on the landmass, such as additional rock material or an accumulation of ice, will tend to depress the land mass. Figure 19 will help to make this clear.

The hypothesis outlined above must involve some compensatory lateral flow of material at depth in the mantle—that is, there must be a lateral flow from beneath the sinking column to

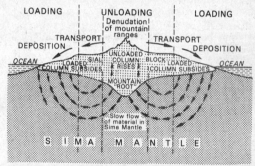

(a) ISOSTATIC EQUILIBRIUM IN THE CRUST *After Airy*

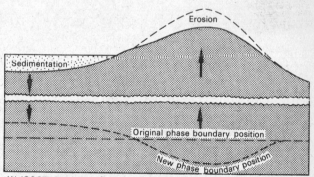

(b) ISOSTATIC MOVEMENT RESULTING FROM PHASE BOUNDARY CHANGES

FIG. 19.—*(a)* The old idea of isostatic equilibrium in the crust, *i.e.* that surface loading and unloading produced, respectively, subsidence and uplift. *(b)* Isostatic movement resulting from phase boundary changes.

under the rising column. Some geologists find little evidence to
support this and have thus suggested an alternative hypothesis
to explain vertical movements. This second theory involves the
concept of *phase change*. It is known that when some materials
are subjected to considerable pressure they will change their
atomic structure and crystal state and compact themselves into
a smaller space—that is, they undergo a phase change. It is
thought that a mineral such as olivine which forms a large proportion of the mantle material will behave in this way. Thus at
a certain level in the mantle where a critical pressure is reached a
phase change occurs and a phase change boundary line can be

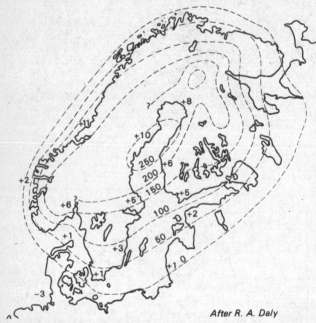

After R. A. Daly

Fig. 20.—*Post-glacial uplift in the Fenno-scandian region.* The weight
of the accumulated ice depressed the land but when the ice melted away
the land began to rise again. The pecked lines show the amount of uplift
in feet in different parts of the area. The figures having positive and
negative values show the rise or fall in sea level in millimetres per year
along different parts of the coast.

drawn in (Fig. 19(*b*)). If the pressure on an area of the earth's
crust is decreased (through erosion for example) then the phase
change boundary line must move downwards in position (*i.e.*
the critical pressure for change is not attained until a greater
depth is reached). This means that some of the mantle material
is changed from a high density to a low density form and consequently increases in volume. This in turn causes an uplift of
the rock column above it. The converse is also true. Loading
(*e.g.* by ice or sedimentation) at the earth's surface would
produce an upward movement of the phase change boundary,
a compaction of some of the mantle material, and a resultant
sinking of the rock column above.

The idea of isostatic readjustment may be illustrated by the
after-effects of the Pleistocene glaciation. The formation of the
great ice sheet over Scandinavia led to the land being depressed
by the enormous weight of the accumulated ice. When, eventually, the ice began to melt, the land surface rose in sympathy with
the gradually easing load. The melting of the Scandinavian ice
resulting in an uplift of the land which, it has been calculated,
reached 900 feet (274 m) in places. This uplift is illustrated by
the occurrence of a series of raised beaches. The uplift is, in
fact, continuing and can be illustrated by changes which have
gone on along the Bothnian coast of the Baltic; Figure 20 shows
the rise or fall of harbours in millimetres per year in this part of
Europe. The Finnish port of Vaasa, for example, has been in
existence for several hundreds of years, but the modern port
lies some six miles (approx. ten km) west of the original harbour
site—a fact which illustrates how the continuing uplift along
this coast has rendered some of the old harbours quite useless.

FOLDS AND FOLDING

5. The effects of compressive forces. The effects of compressive
forces in the earth's crust necessarily involve a shortening or
contraction of the area concerned. That layers of rock are
capable of being crumpled or buckled into folds is a matter of
simple observation: the folds can often be seen in the faces of
cliffs or railway cuttings. The ease with which rocks are folded
and the degree to which the rock layers are crumpled are
dependent upon:

(*a*) the physical characteristics of the rocks involved in the folding, *e.g.* whether they are layered or not, their degree of resistance, etc.;

(*b*) the temperature conditions and the presence of pore fluids;

(*c*) the intensity and duration of the compressive movement.

6. Types of folds. A sequence of fold structures (illustrated in Fig. 21) may be formulated which are related to the degree of compressional forces involved:

(*a*) an *anticline* is formed when the strata is bent upwards into a simple symmetrical linear upfold;

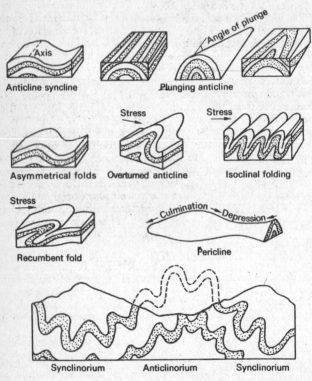

FIG. 21.—*Fold structures.*

(b) a *syncline* results from the strata being bent downwards in a symmetrical manner;

(c) a *monocline* results when horizontally-laid beds dip and then flatten out again, producing a simple flexure;

NOTE: monoclines are often associated with vertical movements (faulting).

(d) an *asymmetrical fold* is where one limb in a fold structure is steeper than the other; asymmetrical folding is characteristic of the chalk downland of the Isle of Wight;

(e) an *overturned anticline* occurs when an asymmetrical or inclined upfold is pushed right over to form an overfold;

(f) an *isoclinal fold* results from the continued lateral compression upon an overfold, crowding it upon the adjacent overfold;

(g) a *recumbent fold*, literally a fold lying down, results from the continuation of pressure; the axial plane and both limbs of a fold lie roughly horizontal; and

(h) a *nappe* results when the pressure exerted upon a recumbent fold is sufficiently great to cause it to be torn from its roots and to be thrust forward. These structures are common in the Alps. (Fig. 22).

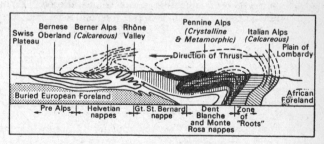

FIG. 22.—*Cross-section of the Alps.* A simplified structural section from north to south across the Alps showing the formation of nappes.

7. Folding and fold mountains. Fold (or folded) mountains are due to forces of compression. Lateral pressure buckles accumulations of sedimentary deposits into parallel upfolded ridges (anticlines) and downfolded troughs (synclines). Their formation has, it appears, occurred in two main ways:

(a) by two earth blocks moving towards one another, or
(b) by one earth block moving towards a stable block.

In either case, the sediments which have been deposited and accumulated in between the blocks will be compressed, buckled and raised up. Fold mountains occupy the sites of former *geosynclines*—great elongated troughs or depressions in which sedimentary deposits of great thicknesses—often as much as 20,000 ft (6,000 m)—gradually accumulated. As the geosynclines filled up, their floors subsided under the weight of the accumulating deposits, thereby making it possible for enormous thicknesses of sedimentary rocks to be built up. Ultimately, the compressive forces squeezed these sediments into gigantic folds to create mountains.

Fold mountain systems are usually extremely complicated in their structure, having various kinds of folds, such as recumbent folds and nappes. All the great mountain ranges of the earth—the Alps, Himalayas, Andes, Rockies—are folded mountains. They are usually termed Young Fold mountains to distinguish them from mountains which, though originally formed in the same way, are much older and which, as a consequence, have been much worn down and altered. Young fold mountains are high, rugged and pyramidal-peaked; they occur in long loops, arcs or festoons, periodically converging to form "knots".

FRACTURES

8. Joints. Fractures or dislocations in the earth's crust are of two main kinds: joints and faults.

A joint may be defined as a crack or plane of fracture dividing a hitherto continuous rock mass into two parts. While some types of rock give or bend under pressure, hard, brittle rocks are more likely to break or fracture.

Joints differ from faults in that little or no actual displacement occurs along them.

Many rocks, *e.g.* limestone, sandstone, granite, have joints in them although they are not necessarily caused by tectonic stresses and strains, for example:

(*a*) in sedimentary rocks the joints may be due to shrinkage resulting from slow drying out,

(*b*) in igneous rocks the jointing is usually caused by contraction on cooling; the cooling of basalt often produces hexagonal columns as in the Giant's Causeway, County Antrim, Northern Ireland.

9. Definition of faulting. A fault may be defined as a fracture which involves the displacement of the rocks on either side of it relative to one another. In faulting there is a movement of greater or smaller extent along the *plane of fracture*; such a movement may be vertical or horizontal or a combination of each.

Any fault structure has a number of characteristic features (*see* Fig. 23) and these are described by a specific terminology:

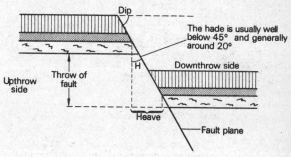

Fig. 23.—*Features of a normal fault.*

(*a*) *throw* refers to the vertical displacement of the strata, which may vary from a fraction of an inch to hundreds of feet;

(*b*) *heave* is the lateral displacement which occurs when the fault-plane is inclined;

(*c*) *dip* is the inclination or angle between the fault-plane and the horizontal;

(*d*) *hade* is the angle of inclination of the plane of a fault measured with reference to the vertical.

10. Types of faults. Fault structures are of various kinds and it is possible to classify them according to the movements which have taken place along them (Fig. 24).

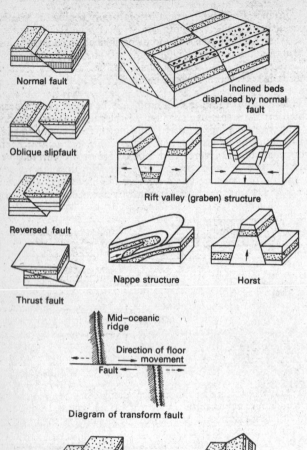

Normal fault

Inclined beds displaced by normal fault

Oblique slipfault

Rift valley (graben) structure

Reversed fault

Nappe structure

Horst

Thrust fault

Mid-oceanic ridge

Direction of floor movement

Fault

Diagram of transform fault

Wrench (tear or transcurrent) fault

Pivotal (rotational) fault

FIG. 24.—*Types of fault and fault structures. See* **10**(*d*) *for an explanation of transform faults.*

(a) *Normal faults*. These are faults which have both the hading and the down-throw in the same direction. In these instances the rock mass moves down the dip of the fault-plane. Normal faults are produced by tension and indicate a pulling apart of the strata or rock mass.

(b) *Reverse faults or thrust faults*. Where compression occurs the beds on one side of the fault-plane may shear and be thrust over those on the other side; in other words, there is over-thrusting up the dip, and the fault hades to the up-throw side. Thrust faulting is common in highly-folded areas such as the Alps.

(c) *Tear faults or wrench faults*. These occur when there is a movement in a horizontal direction although the fracture is vertical. These are sometimes termed transcurrent faults. The Great Glen fault in Scotland is a wrench fault which has a considerable horizontal displacement. Previously this displacement was said to be of the order of 65 miles (approx. 100 km), but a dispute has arisen about the direction of displacement.

(d) *Transform faults*. These are a type of fault largely found in the ocean floor. They cut across and offset the rift zones of the mid-oceanic ridges. Superficially they appear to be wrench or transcurrent faults which have effected a lateral displacement of the ridges. Seismic data and a consideration of sea floor spreading, however, show that the movement is actually occurring in a reverse direction to that apparently shown by the shift of the oceanic ridge.

11. Faulting and block mountains. Earth movements produce excessive stresses and strains in the crust, which result in fracturing. Fracturing is usually accompanied by dislocation. Great masses of the crust may be either reared above, or dropped below, the general level of the land. Such relatively rapid vertical movements produce crust blocks (*horsts*) margined by steep fault-scarps, rift valleys or depressions (*graben*) bounded by faults (*see* Fig. 24). Horsts and graben are found in association with one another. The mountain remnants of the Hercynian mountain-building period of Permo-Carboniferous times in Central Europe, *i.e.* the Central Plateau of France, Vosges, Black Forest and Bohemian massifs, are Block Mountains (they are, in fact, the relics of old fold mountains which have been denuded and shattered). The Rhine Rift Valley forms a graben.

The Flinders Mountains and the Torrens Valley in South Australia are other examples of uplifted and downthrown areas. The relief of dislocated mountains such as we have just mentioned is much more subdued than that of young fold mountains; tops are blunted, squared-off and lowered, and denudation often modifies the steepness of the scarp slopes.

THE EARTH'S MAJOR STRUCTURAL UNITS

12. Continents. When we study the structure and relief of the earth's surface as it appears today, a number of large scale structural features immediately become apparent. To begin with, it is clear that we can differentiate between the continental and continental shelf areas, and the deep ocean basins. The former consist of rocks of granitic composition, some of which may date back to Pre-Cambrian times, whereas the latter are formed of basaltic rock which is nowhere older than Mesozoic age. When we inspect the structure of the continents more closely it can be seen that they contain great patches of very ancient Pre-Cambrian igneous and metamorphic rock. These areas are termed as *continental shields* or *kratons* (*e.g.* the Baltic, Siberian and Canadian shields) parts of which constitute the *nuclei* around which the continents have grown. Surrounding the shield rocks are groups of younger strata which were mostly added during *orogenic movements* or mountain building phases. They take the form of linear fold belts that were produced during three major phases. The first two of these, the Caledonian and Hercynian orogenies, took place around 400 million and 280 million years ago respectively. The mountains which were formed then have undergone considerable subsequent denudation and modification. Sometimes termed the older mountains (*e.g.* Appalachians of North America, East Australian Highlands, etc.), they now rarely exceed 6,000 ft (1,828 m) and are in the main broken into fault blocks with dissected plateau surfaces. The third major orogenic phase took place only about 30 million years ago and indeed we may still be experiencing its effects. This led to the formation of the young fold mountains (*e.g.* the Alps, Himalayas, Rockies, etc.). Such mountains are very high and rugged although some of them embrace broad intermontane plateaus. In some instances (*e.g.* the Himalayas) they are made up almost entirely of sedimentary rocks, whilst in others (*e.g.* the Andes) there is much associated vulcanicity.

Earthquake activity however, is common in the young fold mountains, and in distinct contrast to the older areas they are among the earth's present unstable belts.

The great lowlands of the continental masses include many structurally depressed regions and others which are underlain by horizontally-bedded rocks that have been little disturbed by mountain-building movements (*e.g.* the Great Russian Platform and the larger part of the central plains of North America). Other lowlands are alluvial in character and many are still being extended by the depositional action of rivers and sea-waves. Such are the broad alluvial lowlands of Asia (*e.g.* the Indo-Gangetic Lowland, the Great Plain of North China and the Gulf-Atlantic Plain of the United States).

13. Ocean basins. The great continental shelf areas are structurally part of the continental masses and it is only beyond them on the deep ocean floors that we find a different structural and topographical style. The ocean floors largely consist of comparatively flat *abyssal plains*. Individual hills projecting above the general level are called *sea mounts* or, if flat-topped, *guyots*. They are believed to be of volcanic origin. Yet the oceans are not without mountain ranges and deep valleys. Running across the oceans are a chain of submarine mountains that extend a total distance of over 60,000 km. Such mountain chains are between 1,000 and 1,500 km wide and often rise 3,000 m from the sea floor. They are referred to as *oceanic ridges* (*e.g.* mid-Atlantic Ridge) and are largely formed of basaltic rocks resulting from igneous activity. The ridges have a deep cleft or *median rift* running along their central axes; in places this is over 2,500 m deep. There is no doubt that the vulcanicity, which produces the ridges, is associated with this central fracture zone.

In some oceanic areas the continental shelf and slope is replaced by an elongated *ocean trench* and sometimes an *associated island arc*. Deep sea trenches are very deep linear troughs which lie roughly parallel to adjacent coast lines (*e.g.* Challenger Trench around 11,000 m deep). They contain remarkably little sediment and are zones of considerable seismic disturbance. Located on the continental side of some trenches may be a festoon of volcanic islands (*e.g.* Japan, East Indies, etc.). The consistent eruptions and earthquakes experi-

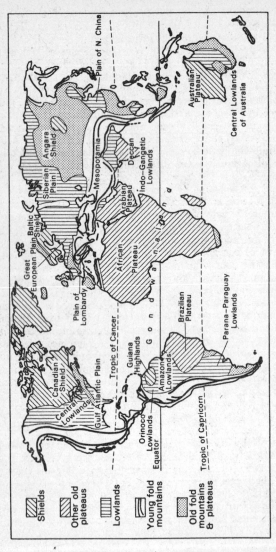

Fig. 25.—The main structural features of the earth.

enced in such zones place them among the world's most unstable belts.

Almost all of these major structural features described above can be explained by the theory of plate tectonics (*see* x, 8). Figure 25 shows in simplified form the elements of structure of the earth's surface.

PROGRESS TEST 8

1. Draw a distinction between the tectonic and denudational forces which are involved in a landscape formation. (1)

2. What is meant by the phrase "orders of relief"? How many orders are there and what scale of magnitude does each order refer to? (2)

3. Explain fully the meaning of the term diastrophism. (3)

4. Write an account of the concept of isostasy. (4)

5. The geologist recognises different types of fold structures; name and briefly describe the various types. (6)

6. What are fold mountains and how may they have been formed? (7)

7. Draw an annotated diagram of a fault structure to show its characteristic features. (9)

8. Draw a distinction between normal, reverse and tear faults. (10)

9. Briefly describe the earth's major structural units. (12 and 13)

10 Explain the meaning of the following terms: epeirogenic movements, geosynclines; horsts, kratons, graben. (3, 7, 11, 12)

CHAPTER IX

EARTHQUAKES AND VULCANISM

EARTHQUAKES

1. The phenomena of earthquakes. Earthquakes are tremors or vibrations in the earth's crust. The tremors vary greatly in their intensity and effects: some are so slight as to pass unnoticed while others may be very severe and catastrophic in their effects. A scale of earthquake intensity was devised a century ago by Rossi; this was subsequently modified by Forel. The scale now used is as follows:

Rossi-Forel scale of seismic intensity

  (1) Noticed only by an experienced observer.
  (2) Noticed by a few people at rest.
  (3) Generally felt by people at rest.
  (4) Felt by people in motion; doors and windows rattle.
  (5) Felt generally; furniture disturbed.
  (6) Sleepers awakened; hanging objects, such as chandaliers, set in motion.
  (7) Causes panic; moveable objects overthrown; church bells ring.
  (8) Damage caused to buildings; chimneys fall and walls are cracked.
  (9) Some buildings collapse and are destroyed.
  (10) Widespread destruction.

Seismic tremors are much more common than most people realise and on the average about four are registered every day although these are slight, of little consequence, and may go unnoticed. However, once every three weeks or so, shocks of some intensity occur, while in any one year, on the average, several severe earthquakes take place which usually result in considerable damage and some deaths. Periodically, catastrophic earthquakes occur which may kill or injure hundreds of people, destroy villages and towns, and render thousands homeless.

2. Seismic recording. The vibrations which are caused by earthquakes can be detected and recorded by an instrument termed a *seismometer*. A seismometer is an instrument which is designed to measure the displacement of the ground with respect to a mass. Figure 26 shows one type of seismometer. In this

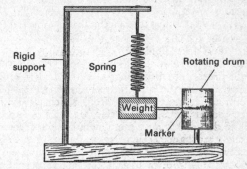

PRINCIPLES OF A SEISMOMETER

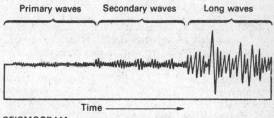

SEISMOGRAM

FIG. 26.—*Seismometer and seismogram.* (*a*) illustrates the principle behind the seismometer (sometimes called a seismograph). A seismometer is merely an instrument which measures the displacement of the ground with respect to a mass. An earth tremor will agitate the spring causing the weight to move up and down; as it does so, the attached pointer registers its vertical movements on the rotating drum. (*b*) is a seismogram; it shows the different types of waves generated by an earthquake and indicates the time of their arrival at a station.

particular instrument a mass or weight is suspended from a horizontal bar by means of a spring. When a tremor occurs, the spring is agitated and this causes the weight to move up and down; as it does so, the attached pointer registers the weight's vertical movements upon the rotating drum. Another type of seismometer measures horizontal movements.

The lower diagram in Fig. 26 is a seismogram or the recording of the vibrations registered by a seismometer. It will be seen that three different types of oscillation are shown on the seismogram: these are known as P, S and L waves. The P and S waves are those that travel *through* the earth; since the P waves have the greater velocity of the two, they arrive at the recording station first and accordingly are registered first; the S waves follow. The L waves, waves of long period, travel around the earth in the crust; they take longest to arrive at the recording station and hence are the last to be recorded on the seismogram.

The point in the earth's crust from which an earthquake disturbance emanates is termed the *focus* of the earthquake. The point on the earth's surface immediately above the point of origin of the disturbance is called the *epicentre* (Fig. 27). From the intensity of the tremors felt or from the amount of damage done, it is possible to draw lines, after the manner of contour lines, which link together all points of equal intensity; such lines are called *isoseismal lines* and it is possible to draw isoseismal maps.

3. Distribution of earthquakes. Tremors are felt over large areas of the earth's crust, although they do not occur everywhere, neither do they occur with equal intensity. There are some areas, *e.g.* the old, stable shield areas and the old plateau masses, that are free from, or virtually free from, earthquake disturbances. On the other hand, there are some areas which are especially prone to earthquake disturbances, *e.g.* the edges of the earth's "plates" (*see* Chapter X). The earth's susceptibility to tremors may be correlated with the great zones of crustal weakness:

    (*a*) The belts of young fold mountains.
    (*b*) The zones of major faulting and fracturing.
    (*c*) The belts of mid-ocean ridges.

A map showing the principal zones of seismic disturbance shows three belts:

    (*a*) The mid-Old World belt from Spain to China.
    (*b*) The circum-Pacific belt which margins the Pacific Ocean.
    (*c*) The mid-ocean ridges, such as the mid-Atlantic ridge.

There is one other important area to which attention should be drawn: this is the Great Rift Valley of East Africa which runs via the Red Sea and the lakes of East Africa. This is a zone of large-scale faulting.

4. The causes of earthquakes. What causes earthquakes? Mostly they result from sudden movements within the earth's crust. Stresses and strains may build up within the crust and when a certain point of intensity is reached fractures or slips may

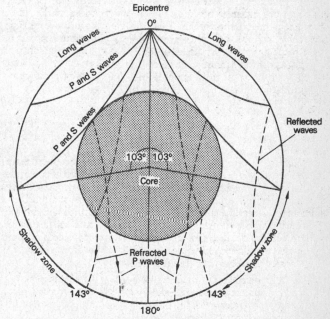

Fig. 27.—*Earthquake waves*. The figure shows the different paths taken by earthquake waves through the earth. The primary or longitudinal (P) waves and the secondary or transverse (S) waves travel *through* the earth; the surface (L) waves travel *around* the earth. Note that S waves do not penetrate the core; also, that the P waves become refracted or bent as they travel through the core. Note, too, the shadow zones between 103° and 143° distant from the epicentre where no P or S waves are received.

occur. Such movements give rise to local crustal jolts, the effects of which are to send out vibrations or tremors. The majority of earthquakes appear to be the outcome of the slipping and settling down of rock masses along fracture lines, although some tremors are associated with vulcanism (but the two are not necessarily related). The stresses and strains set up by mountain building are likely to result in much earthquake activity while another cause may be related to intense surface loading which also causes crustal strains.

Earthquakes are, perhaps, the most horrifying of physical catastrophes mainly because of the suddenness with which they happen and the terrible destructiveness they may bring, but they should be viewed in their right perspective; as Beckinsale has written: "Earthquakes are not major geological processes; they are merely the outward sign of them . . . geologically speaking, their total effect upon scenery is temporary and insignificant" (*Land, Air and Ocean*, Duckworth, 2nd ed. 1965, p. 186).

VULCANICITY

5. Introduction. In the popular and more narrowly interpreted sense, the term vulcanicity is used to describe volcanoes and volcanic activity, but more properly vulcanicity embraces all the activities and processes by which gaseous, liquid and solid substances of internal origin are injected into the crust of the earth or ejected on to its surface. In addition to the spectacular activity of volcanic eruptions there is a great deal of quiet and unseen activity taking place which is just as important.

Basically there are two main processes:

(*a*) The intrusion of masses of magma—magma is a general term for molten rock material—into the upper crust; such rocks which are injected into the crust are termed, as we have already mentioned, *intrusive rocks*.

(*b*) The extrusion on to the surface of molten rock material; this may be simply poured out on to the surface or forcibly ejected through explosive volcanic activity; such material solidifies to form *extrusive rocks*.

6. Intrusive processes. The geologist recognises various intrusive features or forms. These forms, which result from the injection, cooling and solidifying of molten magma beneath the surface, depend mainly upon the degree of fluidity of the magmatic material and the structure and character of the crustal rocks. If the magma is very mobile it will flow more easily and penetrate further than if it is very viscous. Moreover, if the rock has many weak bedding planes or has numerous joints or faults in it, then the intrusion of magma will be facilitated.

Many kinds of intrusive forms are distinguished, chief of which are the following (Fig. 28):

(a) *Batholiths:* these are very large, deep-seated intrusions of plutonic rock. Since, at the time of their formation, they lay deep within the crust, the magmatic material cooled very slowly with the result that the rocks are coarsely crystalline. In due time, the overlying rocks may be eroded away to expose the batholith, *e.g.* the Idaho Batholith.

(b) *Bosses or stocks:* these are similar, but smaller, intrusions. If the intrusions are circular, they are commonly called bosses, if irregular, stocks. If the area of outcrop covers more than about 40 sq. miles (approx. 100 km²) the feature is usually termed a batholith, if less, a stock. The granite mass of Dartmoor is commonly called a stock, although the stock may be a protrusion from a deeper batholith.

(c) *Laccoliths:* these are formed when magma forces its way into overlying strata which are bent upwards to form a dome; the Henry Mountains of Utah, in the United States, provide the classic example.

(d) *Lopoliths:* these are similar lenticular intrusions to laccoliths except that they form saucer-shaped bodies, *i.e.* instead of having upward doming they exhibit concave or sagging floors.

(e) *Phacoliths:* these are also lenticular intrusions of igneous rocks which have been forced into and lie near the crest of an anticline or the base of a syncline in folded strata.

(f) *Sills:* where magma finds its way along bedding planes to form more or less horizontal sheets of igneous rock, sills are formed. Fair Head Sill, which outcrops on the coast of Antrim in Northern Ireland, provides a spectacular dolerite sill, 250 ft (76 m) thick.

(g) *Dykes:* where vertical intrusions cut across bedding planes, probably taking advantage of a joint or fault in the

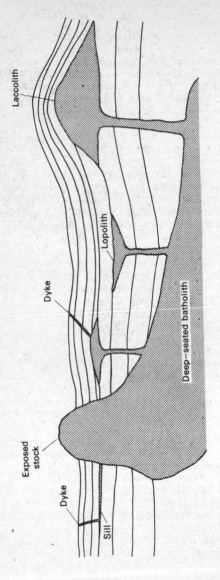

FIG. 28.—*Types of intrusion.*

rock, dykes are created; whereas a sill is concordant with the strata, a dyke is discordant with it.

NOTE: some of these intrusive forms are discordant, *i.e.* they cut across rock strata, as in the case of batholiths, bosses and dykes, while others are concordant, *i.e.* they are lenticular, such as laccoliths, lopoliths, phacoliths and sills.

7. Volcanic Eruption.

Volcanic eruptions are of two principal types:

(*a*) Linear or fissure eruptions, *i.e.* when the magma wells up and pours out on to the surface through a crack or fissure in the crust.

(*b*) Central or vent eruptions, *i.e.* when magma is forcibly ejected through a localised vent hole in the crust.

Which of these two types occurs depends upon two main factors, the nature of the exit outlet, and the composition of the magma.

8. Fissure eruptions.

These form the simplest type of eruption. Usually the lava wells up from below, escapes on to the land surface via a crack or fissure, and pours out quietly on to the surface. Typically there is little or no explosive activity. The molten material flows easily and covers the existing land surface with a blanket of basalt. The lava may reach the surface along the entire length of a fissure or it may pour out more erratically at a series of points. Fissure eruptions commonly take place intermittently in a series of successive flows over a fairly long period of time. Individual outpourings are not usually very thick; on a rough average they attain a thickness of around 20 ft (approx. 6 m). The total accumulated thickness of numerous individual outpourings may, however, be very great and some of the lava sheets are thousands of feet thick.

At the present time few fissure eruptions are occurring on the land surface, although they appear to be taking place on the ocean floors where there are mid-oceanic fissures separating "plates," *e.g.* along the mid-Atlantic Ocean ridge. There is, however, plenty of evidence to show that in former geological ages fissure eruptions took place on an extensive scale in various parts of the world, for example:

(*a*) The Deccan Plateau in the Indian sub-continent has a vast basalt blanket which covers some 400,000 sq. miles (approx. one million km²) and has a general thickness of some 4,000 to 6,000 ft (1,219–1,829 m).

(b) The Columbia-Snake Plateau in the north-western part of the United States covers quarter of a million sq. miles (646 million km²) and has a maximum thickness of about 5,000 ft (1524 m).

(c) The Parana Plateau in South America which covers part of southern Brazil and Paraguay is another extensive basalt plateau believed to be the result of fissure eruptions.

9. Central eruptions or volcanoes. More common are volcanoes which result from central eruptions. In the case of volcanoes activity is localised, the lava pouring out from a single orifice (Fig. 29). When this happens the extruded material builds itself up to form a cone, a shape commonly associated with the

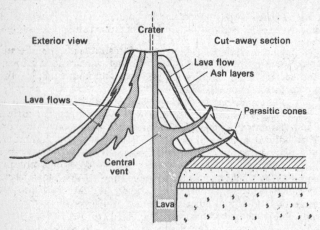

FIG. 29.—*Cross section of a volcano.* External and interior view of a volcano built up of layers of ash with occasional lava flows.

popular concept of a volcano. However, it should be emphasised that central eruptions do not always give rise to cone-shaped mounds; there are cases, as in central Italy, where ejections of ash have blanketed the land surface and not produced cones.

Occasionally volcanoes are created as a result of a single phase of vigorous activity, *e.g.* Monte Nuovo in Italy, which erupted in 1538, formed a cone some 440 ft (135 m) high within three days; but usually volcanoes are built up as a result of a series of intermittent phases of activity. Long-continued activity, perhaps over several hundreds of years, is likely to produce very large cones which may reach many thousands of feet in elevation, such as some of the Andean volcanoes or those of South east Asia.

Volcanoes may be said to go through four main phases in their life cycle:

(*a*) *The active phase* when they show signs of activity; active volcanoes may be:

(*i*) constantly eruptive;
(*ii*) intermittently eruptive.

(*b*) *The dormant phase* when they appear to have exhausted themselves and remain quiet for long periods but may suddenly burst into renewed activity.

(*c*) *The extinct phase* when they have not erupted for several thousand years and are believed to be incapable of any further activity.

(*d*) *The destructive phase* when the volcano undergoes pronounced erosion and is partially or substantially destroyed, perhaps leaving behind only remnant plug.

10. Extrusive materials. The materials that are extruded on to the surface are the following:

(*a*) *Gases.* Carbon dioxide, sulphur dioxide, hydrogen sulphide, chlorine, fluorine and boron may all be emitted. Most of them contain a high proportion of super-heated steam and it is, in fact, this association of gas and steam which is largely responsible for the explosive characteristics of volcanoes. Sometimes masses of gas, super-heated steam and incandescent dust roll down the slopes of volcanoes during periods of eruption; these clouds of glowing matter are termed *nuées ardentes*. These are poisonous and highly destructive.

(*b*) *Water.* Some steam is normally produced when volcanoes are in their active state. If steam is emitted in quantity it may give rise to heavy downpours when it becomes cooled and condensation occurs. Torrential rains are, in fact, a frequent accompaniment of volcanic eruptions. If the rain mixes with fine volcanic dust, mud-flows (lahars) may result which may swamp and obliterate settlements as happened in the case of Herculaneum, a Roman town not far from present-day Naples. Nearby Pompeii seems only to have been smothered in ash.

(c) *Lava*. Lava is the general name given to molten rock. Rock becomes "liquefied" within the crust or upper mantle as a result of the combined action of temperature and pressure. The lava finds its way to the surface and then flows from the crater downslope. The speed at which lava flows depends, in part, upon declivity but also, and to a very considerable extent, upon its composition and temperature: if the lava is basic it is very mobile but if it is acid it is viscous. The degree of fluidity of lava has an important bearing upon the form which a volcanic cone takes—whether the slopes are gentle or steep.

(d) *Solid matter*. Most volcanoes eject a certain amount of debris in solid form. Six main kinds of solid debris, known as pyroclasts or tephra may be distinguished:

(i) *Volcanic breccia:* this consists of angular fragments of volcanic rock which are formed when the solidified lava in the volcano crater and pipe is shattered at the onset of activity.

(ii) *Pumice:* this is the solidified scum from the surface of the lava; it has a spongy or cellular character due to the bubbles of steam and gas in the lava scum when it cooled.

(iii) *Scoria:* this is ejected material of a cindery or slaggy nature.

(iv) *Lapilli:* these are small round or angular fragments of rock, varying in size from about $\frac{1}{4}$ to 1 inch (6—25 mm) diameter, which are thrown out during explosive phases.

(v) *Volcanic bombs:* these are clots of lava of roughly rounded shape but of variable size which are ejected from the crater.

(vi) *Tuff:* this is volcanic ash or dust, with particles less than 4 mm in size, which accumulates and becomes compacted and cemented into consolidated masses of rock.

11. Classification of volcanoes. Volcanoes can be classified in three main ways:

(a) According to the degree of activity (this we mentioned in 9 above).

(b) According to the explosive nature of the volcanic material.

(c) According to the composition of the volcanic material.

The following classification is according to (b). The six types are recognised on the basis of the character of the volcanic explosion which is itself dependent upon the pressure and quantity of the gas present in the magma. The range of explosive activity may be illustrated by type example, as follows:

(i) *The Hawaiian type.* In this case gases are liberated quietly and there is a general absence of explosive activity. Lava wells up from below and, because it is basic in composition, it is very mobile. Because the lava flows readily, it travels far before solidifying and so tends to build up cones of low angle of slope.

(ii) *The Strombolian type.* Named after the volcano of Stromboli, near Sicily, the explosive activity is mild in in character but occurs with a considerable degree of regularity at short intervals. With every explosion red-hot clots of lava are ejected which form "bombs" of scoria.

(iii) *The Vulcanian type.* Named after Vulcano in the Lipari Islands, the explosions in this type are of a more violent character and are accompanied by dark clouds and much steam. The explosions, however, occur at more irregular intervals. Since the lava is more viscous, it tends to crust over and so "bottles up" the gases until they explode with considerable violence.

(iv) *The Vesuvian type.* Vesuvius, near Naples, in Italy provides the type example. In this case, long periods of quiescence or very subdued activity are shattered by violent explosions. Gases build up in the magma chamber below and become highly explosive; eventually the pent-up gases explode with great force, shattering the rock which has plugged the vent and expelling the lava.

(v) *The Krakatoan type.* So called after the extreme explosive violence associated with the eruption of Krakatoa in the East Indies. Here the explosion was so tremendous that the volcano itself and much of the island was blasted away. In this case, although vast quantities of volcanic dust were ejected, there was no lava emitted.

(vi) *The Peléan type.* This is the type associated with

the explosive eruption of Mont Pelée, in Martinique, in the West Indies. The eruption, which occurred in 1902, was characterised by blasts of dark or incandescent gas and ash (*nuées ardentes*) which, unable to escape upwards because the vent was sealed by a lava dome, were discharged through lateral cracks in the volcano. Twenty thousand people lost their lives and the town of St. Pierre was destroyed.

12. Classification according to composition.

(a) *Shield volcanoes* or *lava shields*. If the lava is basic in its chemical composition, it tends to be very fluid or mobile and will flow for considerable distances before it cools and solidifies. Such fluid lava results in the building up of cones which are broader than they are high and which have small angles of slope. Such laval cones are known as shield volcanoes; the volcanoes of Hawaii are of this type. Typically, they are of vast size.

(b) *Dome volcanoes* or *cumulo domes*. On the other hand, if the lava is acid it is rather stiff or viscous and does not flow readily. Such lava cools and congeals quickly and thus does not flow very far. Hence the lava tends to pile up and produce cones which have an elevation which is great in proportion to their basal diameter; moreover, the cone tends to assume a dome shape, although this distinctive shape seems to be due in part to internal pressure and expansion. Typical examples of this kind of volcano occur in the Auvergne region of central France.

(c) *Cinder cones*. When volcanoes undergo explosive activity showers of volcanic debris—ash, cinders, lapilli, bombs, etc.—are ejected into the air; this fragmental material falls back to earth and gradually builds up a cone. Such cinder cones are steep-sided but rarely reach any great height, seldom more than 2,000 ft (600 m) although occasionally, as in the case of Volcano de Fuego, in Central America, (11,000 ft; 3,352 m) they may attain exceptional elevations. Sometimes they are merely explosion holes.

(d) *Composite cones* or *strato-volcanic cones*. These are cones which are built of both fragmental material and lava; phases of lava flow alternate with the explosive ejection of ash, etc. Composite volcanoes are the most common form and provide what may be styled the "typical" volcano. They are among

the highest and most imposing of all types of volcano, *e.g.*
Mt Etna, Popocatepetl. Composite cones usually have well-
developed craters and may also exhibit "nesting" (*i.e.* one
inside the other) craters.

13. Distribution of volcanoes. There is abundant evidence to
show that vulcanism has been a feature of the earth's surface
throughout geological time but mostly, and certainly at the
present day, volcanoes are associated with crustal movements
and lines of weakness in the crust (Fig. 30). They occur:

(*a*) close to continental coastlines, *e.g.* along the western
coasts of the Americas;

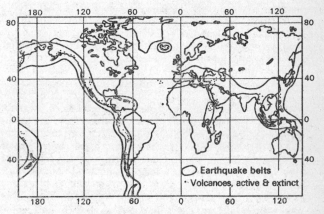

FIG. 30.—*Distribution of volcanoes and seismic zones.*

(*b*) along mid-ocean submarine ridges, *e.g.* the volcanoes of
the mid-Atlantic Ocean ridge;

(*c*) in regions of faulting and earthquake disturbance, *e.g.*
the Middle East and the East African Great Rift Valley;

(*d*) in zones of recent mountain building, *e.g.* the Andes and
the fold mountains of South-east Asia.

A map showing the active land volcanoes, which total

approximately 800, indicates very clearly that there are two main belts:

> (i) the Alpine-Himalayan belt which runs roughly latitudinally across the Old World landmass;
>
> (ii) the Circum-Pacific belt which margins the Pacific Ocean to form "the girdle of fire."

14. Volcanic landscapes. These, though fairly limited in extent, exhibit a considerable variety of land forms, some of a major, others of a minor, nature. Some volcanoes bring a dramatic element into the landscape as, for example, Izalco, in Central America, which is almost continuously aflame. Large, isolated cones such as Mt Etna in Sicily and Mt Egmont in North Island, N.Z., form dominant elements in the landscape. The small, rounded, dome-shaped, craterless volcanoes of the Auvergne in the Central Massif of France give to that area a highly distinctive landscape. In the same region the Mt Plomb du Cantal, a large volcano, over 6,000 ft (1,854 m) high, has had its slopes severely eroded by radial drainage with the result that *planezes* or triangular-shaped plateaus have been formed. In many cases, where the volcano as such has been virtually destroyed by erosive action, only the volcanic plug in the vent of the volcano remains as an upstanding spine, *e.g.* the Rocher de Corneille in central France, Wase Rock in eastern Nigeria. *Calderas* are spectacular landscape features; they are great crater-like hollows which may be anything up to 10 miles (16 km) in diameter. The term means a "basal wreck"; their origin is uncertain; one view holds that they are the remains of volcanoes after paroxysmal explosions shattered and removed the summits of former cones, another suggests that they are due to the collapse of cones which have been weakened (through the tremendous weight they exert on the land surface) and where faulting has taken place since in some cases circular faults bound the depressions. Examples of calderas are Crater Lake, in Oregon, U.S.A. and the crater lake of Öskjuvatn in Iceland. Minor volcanic forms include *geysers*, such as "Old Faithful" in the Yellowstone National Park, U.S.A.; *thermal springs*, such as those of Bath, in England; *mud volcanoes* which are common in North Island, N.Z.; *solfataras*, vents which give off sulphurous gases; and *fumaroles*, vents from which steam is mainly issued. These various minor forms usually mark stages in the final decay of vulcanism.

PROGRESS TEST 9

1. What is the Rossi-Forel Scale and what is its purpose? (1)

2. Describe and account for the distribution of earthquakes. (3)

3. Write brief explanatory notes on the following: seismometer, isoseismal lines, seismogram. (2)

4. Describe the different kinds of intrusive forms distinguished by the geologist. (6)

5. What are the principal distinguishing features between fissure and central eruptions? (7–9)

6. "Extrusive materials are gases, liquids and solids." Elaborate upon this statement. (10)

7. Describe the phases in the life cycle of a volcano. (9)

8. Suggest a classification of volcanoes based upon *either* the explosive activity *or* the composition of the volcanic material. (11, 12)

9. Describe the distribution of volcanoes on the earth's surface. (13)

10. Explain what is meant by the following terms: laccolith, dyke, tuff, lapilli, caldera, geyser. (6, 10, 14)

CHAPTER X

THE ORIGIN OF THE CONTINENTS

DISTRIBUTION OF LAND AND WATER

So far as we know the earth is unique among the planets of the Solar System in that its surface comprises land and water. If the earth was a perfectly homogeneous body, either all land or all sea, then the existence of human life—perhaps all terrestrial life—as we know it would probably be impossible. However, rather less than one-third of the earth's surface is land and rather more than two-thirds water. Two fundamental questions therefore pose themselves:

(a) Do the continental land-masses differ from the ocean floor from which they rise up in their structure and composition?

(b) Can these continental masses undergo uplift or, alternatively, submergence, and have they done so?

The origins of the continents, and of the oceans too, have fascinated man from very early times, as is illustrated by the myth of the lost continent of Atlantis. In this chapter we shall review the theories and the evidence relating to the origin of the continents, although it should be made clear at the start that we still do not know with any certainty precisely how the continents came into being. Let us review, step by step, the theorectical speculations concerning continental origins.

THE THEORIES

1. **The Tetrahedral Theory.** The antipodal arrangement of land and water together with other peculiarities of land distribution over the earth's surface along with the roughly triangular shapes of several of the continents led, in the past, to some theorising about the form of the earth. Lowthian Green, for instance, formulated his Tetrahedral Theory in which he compared the earth to a tetrahedron—a three sided pyramid

with a flat triangular base. The theory assumed that the earth started as a hot body—we are by no means sure of this and some geologists believe the earth has always been cold—and that as it cooled it contracted, the shrinking giving rise to a tetrahedral form since of all regular bodies the tetrahedron has the minimum volume for a given surface. Lowthian Green assumed that the corners of the tetrahedron corresponded to the continents, or more precisely the nuclear shield areas of the continents which are roughly equidistant from one another, while the faces of the tetrahedron corresponded to the oceans (*see* Fig. 31). This

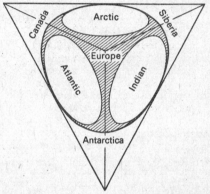

After A. Holmes

FIG. 31.—*Green's tetrahedral concept.* The first of a number of geometrical theories to explain the structure of the earth's surface. In this case the corners of the tetrahedron were thought to represent the "shield" areas, the faces the ocean basins.

theory assumed that the continents have always been stable, in the sense that they have always occupied the same positions on the earth's surface but, this apart, Green's theory has two fundamental weaknesses:

    (*a*) the crust of the earth simply does not have the properties of rigidity necessary to support the idea of a shrinking interior; and

    (*b*) the curious features of land shape and distribution are, in all probability, completely accidental.

It should be remembered that this theory was conceived (1875) before it was known, first, that the continents have not always been stable but have moved their positions throughout geological time, and secondly, that the crust is thicker in continental than in oceanic areas.

2. The theory of drifting continents. The idea that the continental landmasses, instead of being stable, may have moved, *i.e.* suffered lateral displacement, was first suggested as long ago as 1858 by Antonio Snider but at that time the idea seemed so preposterous that it was quickly dismissed and forgotten about. Later, at the beginning of the present century, an American, F. B. Taylor, and an Austrian, Alfred Wegener, both suggested that continental movement may have taken place on a gigantic scale. Both scientists, incidentally, developed their unorthodox ideas quite independently. But it was not until Wegener published his now famous book on the subject, *The Origin of Continents and Oceans*, in 1915, that the idea of continental drifting attracted world-wide attention and began to receive serious consideration.

Later, Wegener's theory was rejected by many geologists, largely because Wegener could not suggest a satisfactory mechanism for causing continental movements to take place, although many others continued to accept Wegener's ideas with reservations. In recent years, largely because of the new evidence supplied by studies of rock magnetism, Wegener's ideas in modified form, have gained a new respectability and much support.

3. The evidence for continental drift. Briefly, according to Wegener, the present day distribution of the continents is due to the breaking up by drifting of an original single major land mass or proto-continent which he termed *Pangaea* (Fig. 32). As a result of this breaking two continental land masses came into existence: a northern block, named *Laurasia* (comprising most of present day North American and Eurasia) and a southern block, styled *Gondwanaland* (including most of South America, Africa, Arabia, the Indian Deccan and Australia). Subsequently, both of these great continental blocks were shattered into smaller fragments and by a process of "drifting" became dispersed and gradually attained their present positions and shapes.

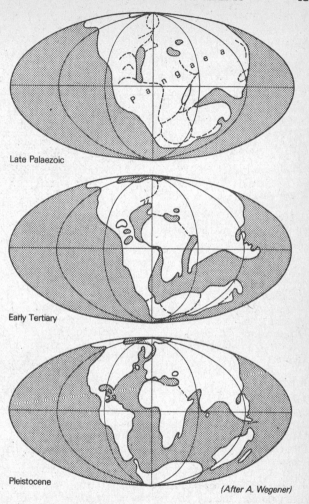

Late Palaezoic

Early Tertiary

Pleistocene

(After A. Wegener)

FIG. 32.—*The evolution of continents according to Wegener*. The figures show how the super-continent of Pangaea slowly broke up into continental fragments and gradually wandered apart.

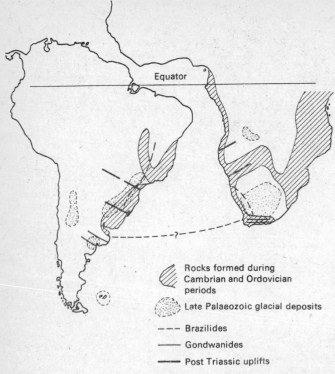

Fig. 33.—*Geological similarities on the two sides of the South Atlantic.*

Wegener and later, his disciple the South African geologist, Du Toit, assembled an impressive amount of evidence to support the idea that the continents had drifted. Some of the main points providing evidence in favour of the theory are as follows (Fig. 33):

(*a*) There is a striking parallelism of the opposing coasts of the Atlantic, and also of some coasts elsewhere in the world, and if they are drawn together they make a rough jig-saw fit. An even better fit is achieved when the edges of the continental shelves are brought into juxtaposition.

(b) There is a close structural resemblance and many geological similarities especially between the eastern coast of South America and the western coast of Africa.

(c) There is evidence from the distribution of certain past plant and animal species, e.g. of marsupials, the Glossopteris flora, and the freshwater worm Phreodrilus, whose occurrence seems quite inexplicable unless continental drifting is invoked.

(d) There are traces of glaciation, dating from the Carboniferous period, found in Australia, India, Africa and South America but with these land areas in their present position the occurrence of the Carboniferous glaciation, which seems to have affected each of them, is very difficult to explain.

(e) Wegener's postulated drifts go far towards offering a reasonable explanation of the creation of fold-mountain systems; they might have been formed by the squeezing, folding and lifting of geosynclinal sediments of the advancing continental blocks.

4. Convection currents. The great plausibility of Wegener's theory of drifting continents foundered very largely in his inability to suggest a satisfactory means of "engineering" continental drift. He talked vaguely of currents in the sub-crust moving the continental blocks but these convinced no one. However, it must be recognised that the cumulative evidence supporting the theory is massive and the theory very attractive. If all the points in the theory could be established and an adequate motive force discovered then, as Professor Shand has said, "we should have to credit Professor Wegener with the greatest piece of geological synthesis that has ever been accomplished."

The search for an adequate motive force led many geologists on to the idea of convection currents within the mantle. The idea of convection currents came to be used to try to explain the fundamental differences in the structure of the continental masses as compared with the crust beneath the oceans. Convection mechanisms have been propounded by many scientists, including the Dutch geophysicist Professor Vening-Meinesz, Professor B. Gutenberg (who discovered the Gutenberg discontinuity between the core and mantle) and the late Professor Arthur Holmes who was professor of geology at Edinburgh University.

Briefly, the idea of convection currents occurring in the mantle is based on the supposition that the interior of the earth, at least in earlier times, was in a hot and liquid state and sufficiently fluid to allow convection currents as may be demonstrated in a beaker of water heated from below by a Bunsen burner. The heat generated to cause the convection was, it was argued, most likely due to the releasing of energy through the breaking down of radioactive elements. In due course, over long ages, as a result of this constant vertical convection, a differentiation of the original material of the earth occurred: the lighter fractions rose to the surface as a scum (sial), which subsequently formed the continents while the heavier fraction sank downwards to form the sima of the ocean floors (*see* Fig. 34).

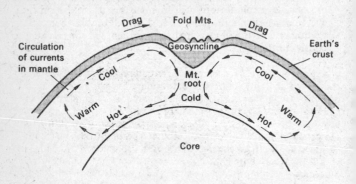

FIG. 34.—*Holmes' convection currents.*

5. The expanding earth. A more recent theory advanced to explain the origin of the continents is that of global expansion, a theory first put forward in 1935 by J. K. E. Halm and subsequently re-stated and elaborated by Bruce C. Heezen. This theory assumes that in the early years of its formation the earth was much smaller than it is now, having a diameter roughly one-half of its present one. As the earth cooled down a crust was formed over its entire surface. Subsequently, as the earth expanded in size, the original crust fractured and its remnants formed "continental rafts," the ancestors of the present continents. Now, if it be assumed that the earth is expanding like an inflating balloon and that the original "continental rafts" have remained roughly the same size, then additional crust would have to be formed. This new crust is alleged to have been formed in the oceans by basaltic rock being forced through from below and coming to the surface along the great cracks which are known to exist in the centres of the Altantic, Indian and Pacific Oceans. As this basaltic material oozed out, it spread out over the expanding earth to form the ocean floor. Hence, as the ocean bed grew, so the continental land masses became further and further separated and, according to this theory, are continuing to do so (see Fig. 35).

This explanation of the origin and location of the continents and oceans is still quite speculative but it provides reasonable explanations for a number of things and overcomes some of the difficulties of Wegener's theory.

6. Palaeomagnetism. Palaeomagnetism may be defined as the study of the earth's magnetism during the geological past. Palaeomagnetic studies reveal the orientation of the magnetic field of rocks at the time when they were originally laid down. This orientation arises in the following way: when sedimentary and igneous rocks are initially laid down or extruded, the magnetic particles within them take on the same direction and dip as the local geomagnetic field *at the time of their consolidation*.

Research during recent years into the magnetic structure of rocks has shown that the magnetic orientation of the fields of the continental rocks varies between one geological age and another; in other words, the "fossil" magnetism provides a means whereby the positions of the magnetic poles of the earth may be pin-pointed at different times during past geological items. From the evidence of palaeomagnetic studies it has been demonstrated that the polarity of the earth's magnetic field periodically reverses.

Surveys of the rocks of the ocean floor in recent years have revealed that they show a zebra-stripe pattern "in which the intensity of magnetisation changes abruptly in linear ribbons parallel to the nearest oceanic ridge." According to F. J. Vine and D. H. Matthews, both of the University of Cambridge, these magnetic patterns provide evidence of:

(a) polar wandering;
(b) sea-floor spreading.

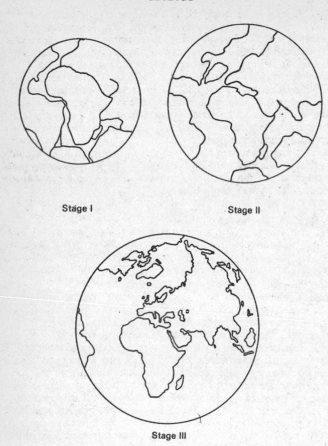

Stage I   Stage II

Stage III

FIG. 35.—*The expanding earth*. Halm suggested the earth might be expanding like a balloon being inflated. Here the increase in volume is exaggerated. Stage I shows the sialic layer, which originally covered the entire surface, just beginning to break up; II shows an expanded earth with the proto-continents separated by embryonic ocean basins; III the earth as it is at the present time.

Thus palaeomagnetic studies are of great importance since, first, they support the theory of continental movements on an appreciable scale and secondly, they support the idea of the continuous creation of new oceanic crust.

7. The concept of sea-floor spreading. Reference has already been made to the occurrence of mid-ocean ridges. These have long been known to exist but oceanic exploration and seismic evidence during more recent decades have shown that they are great zones of crustal weakness. For example, they exhibit linear belts of volcanoes and suffer frequent earthquake disturbance. The concept of sea-floor spreading involves the idea that the ocean floor is, as it were, being continually pulled apart along great cracks or belts of weakness such as we have just mentioned. Molten basaltic material from the depths of the earth is forced upwards through these cracks, helping to prize apart the ocean floor and also to create new ocean crust.

This is an astonishing and fascinating idea but such evidence as we have suggests it is more than just an idea—it is a fact. It has been claimed that the ocean floor has been spreading at the rate of 2–18 centimeters a year. This may seem very small indeed but over a period of some 200 million years the cumulative distance is considerable. If this idea of sea-floor spreading is accepted then it means:

either that the earth is gradually expanding like a balloon as was indicated in **5** above;
or that the earth's surface is being destroyed in some place or places at the same rate at which it is being created along the mid-ocean ridges.

There are many difficulties in accepting the idea of an expanding earth, hence many geologists and geophysicists support the second idea and, as John F. Dewey has said: "Thus there must be, in general terms, a global conveyor-belt system or surface motion that links zones of surface creation and surface destruction."

PLATE TECTONICS

8. The theory of plate tectonics. As we have seen the idea of moving continents goes back over fifty years and while the theory languished because specialists in various branches of science were able to pick holes in parts of the theory, recent studies in palaeomagnetism resurrected the idea. Furthermore, several recent and dramatic discoveries in marine geology have completely revolutionised ideas of the way continents have been formed and mountains created.

The new theory of plate tectonics links together the ideas of sea-floor spreading with the older hypothesis of continental drifting. There are two parts to the new theory:

(a) a *geometric part* which conceives the lithosphere, or outer shell of the earth, to consist of a mosaic of "plates" or stable, rigid segments of crust and upper mantle which are part land and part ocean; in other words, the earth's crust is like an egg-shell which has been cracked in a number of places;

(b) a *kinematic part* (kinematic means relating to motion) which suggests that the various lithospheric segments or plates, large and small, for they vary in size, are in constant relative motion; these plates move on a mobile zone in the upper mantle: this mobile zone is commonly called the asthenosphere.

This new theory of plate tectonics advances the idea that the earth's outer crust is divided up into a number of rigid, shifting plates of varying size—six major ones which are of continental proportions and a number of others which are quite small—and that as these plates slide past one another, converge or move apart, continents drift, mountains are formed, and new crust comes into being.

9. Plates and their margins. The theory of plate tectonics just outlined suggests that the earth's surface is divided up into a number of segments of varying size which are all slowly moving. There are some segments, usually of large size, that are relatively free from earthquake disturbances, and such segments or plates are termed *aseismic plates*. These are usually separated by long, narrow belts where earthquakes are of frequent occurrence and such belts are termed *seismic zones*.

Four differing types of plate margin can be distinguished:

(a) *Constructive or ocean ridge margins.* These are the plate margins adjacent to the great mid-ocean floor ridges with their great cracks or fissures through which magma is poured out. As the plates move apart and as the magma solidifies to form basalt, the plates, through the accretion of basalt along their leeward edge, become enlarged. Hence, such margins are termed constructive plate margins. Examples are the mid-Atlantic margins of the American and African plates.

(b) *Destructive margins*. Just as new ocean floor is being formed in some places, so in others it is being destroyed. For example, as the South Atlantic–South American plate moves westwards it must collide with the Nazca (S.E. Pacific) plate moving eastwards; one leading edge must give way and in this case it is the oceanic crust of the Nazca plate which plunges down beneath the South American plate (*see* Fig. 36). Hence the Nazca plate margin is said to be a destructive margin.

(c) *Destructive margins of intracontinental type*. In some cases two continental plates move towards one another and in so doing destroy the ocean floor between them, as presumably happened when the Australian plate—the Indian sub-continent is part of it—moved towards the Eurasian plate. In this instance, the accumulated sediments on the continental margins were squeezed and uplifted to form the Himalayan system of mountains.

(d) *Conservative margins*. These are the margins where plates slide past each other and where there is, as it were, neutral activity: the plates neither gain nor lose material. It is believed that the great San Andreas fault, along the western margin of North America, which has long been recognised as a line of major seismic activity, marks a sliding zone or line of plate contact.

10. Causes of movement. Wegener's theory of continental drift largely foundered on his inability to suggest a plausible force capable of disrupting his vast proto-continental blocks and capable of moving the continental masses. And it must be admitted that the mechanism required to explain plate movement and consequent plate tectonics is not known. The theory rests on the assumption that some kind of convection process occurs within the mantle.

As mentioned earlier, Arthur Holmes, in 1929, put forward the suggestion that convection currents operated within the mantle. There is today some evidence from a variety of sources

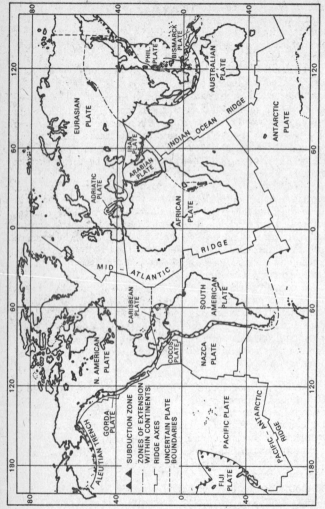

FIG. 36.—*The plates of the earth's crust.* Note the plates are of various sizes and may consist of *continent* and *sea floor*.

which suggests that there *may* be a circulation of hot and less dense material in the mantle: this material rises towards the surface, spreads out, cools and then sinks again. It is suggested that these convection currents carry new material up through the ocean floor ridges, but that this demands that compensatory material is taken in at the ocean trenches. It has been suggested that the active volcanoes which are frequently associated with these trenches may function as outlets for gases released when the lower density rocks of the crust enter the mantle.

An alternative to the idea that convection currents supply the mechanism for movement is that gravity could be responsible. In this case it is suggested that gravity gives rise to push and pull movements. As the leading edge of a plate is subducted or forced under the edge of an adjacent plate, the subduction will have a dragging effect upon the mobile material in the asthenosphere. It is also suggested that as mountains are built on the "lifted" edge of the adjacent plate, the increased weight will exert a downward-pushing effect.

It seems fairly clear that plate movements must be closely associated with the structure of the earth and the character of the mantle. So far the problem of drifting has not been solved and it seems likely that any solution will have to await increased knowledge of the geophysics of the earth's interior.

11. Mountain building. The problem of mountain building has long puzzled geologists. They have, of course, known for a long time that mountains have been built up of vast accumulations of sedimentary deposits which have been squeezed, folded and uplifted, but the precise mechanism whereby this has taken place has long eluded them, although many suggestions have been put forward, *e.g.* the cooling and shrinking of the earth, the advance of continental forelands, etc.

Plate tectonics offers a reasonable and satisfying explanation. Reduced to its basic essentials the idea is that when two plates collide the leading edge of one is subducted or forced under the edge of the other. However, since the continental crust of the second plate is too buoyant to be forced down into the subcrustal zone, mountains are formed along its edge by the crumpling of the marginal rocks and their up-thrusting by the subsiding plate (*see* Fig. 37). For instance, it is believed that the Himalayas may well have been formed when the plate of the Indian Deccan, an ancient rigid plateau block, came into

Stage 1

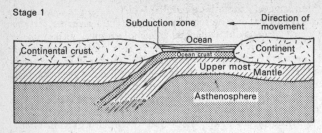

Stage 2

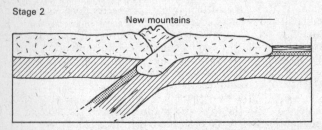

Fig. 37.—*Plates and mountain building.* The diagrams show the stages in mountain building. Stage I shows the approach of a plate; II shows the collision of two plates with the leading edge of one being subducted. Sedimentary deposits on the edges of the plates are crumpled up to form new mountains.

collision with the old Asian plate in mid-Tertiary times; the Indian plate was forced under the Asian plate causing it and the accumulated sediments upon its margin to buckle and be lifted up.

The existence of some mountain ranges, *e.g.* the Urals in the middle of a continental area, does pose certain problems, for such mountains do not appear to fit in with the idea of the collision of plates; but such anomalous ranges can be explained, it is argued, by the fact that continental areas may have divided and become joined together again, perhaps several times, in the geological past and it has been suggested that the Urals may mark the site where a former plate turned down into the mantle.

PROGRESS TEST 10

1. Briefly describe the essential points of Lowthian Green's Tetrahedral Theory. What are the basic weaknesses of this theory? (1)

2. What evidence is there to support the general idea of continental drifting ? (3, 6)

3. Briefly discuss the idea of convection currents in the earth's interior. (4)

4. Write a brief account of the concept of an expanding earth. (5)

5. What is palaeomagnetism and of what value are palaeomagnetic studies to the geologist ? (6)

6. Discuss briefly the concept of sea-floor spreading. (7)

7. What ideas have been put forward to explain mountain building ? (3, 11)

8. What ideas or theories are associated with the following people: Wegener, Halm, Lowthian Green, Holmes ? (1, 2, 4, 5)

9. Explain the following terms: tetrahedron, Gondwanaland, convection currents, polar wandering, plates. (1, 3, 4, 6, 8, 9)

10. "The greatest piece of geological synthesis that has ever been accomplished." Discuss this statement in relation to Wegener's theory of continental drifting. (2, 3, 4)

CHAPTER XI

SURFACE PROCESSES AND WEATHERING

LANDSCAPE CHANGE

1. The changing landscape. The land upon which we live looks solid, unchanging, permanent, but in actual fact it is far from being so. Slowly but surely changes in the landscape take place and the hills, valleys and coasts are altered in their size and shape. The most obvious visible changes are those resulting from catastrophic happenings such as volcanic activity, landslides or storm damage; but, in addition to these more dramatic happenings, the earth's surface is constantly being changed in other more subtle and insidious ways.

The study of such landscape changes and the landforms which result forms the science of *geomorphology*, a study common to both geography and geology.

2. Physical forces of change. The landscape is the outcome of two physical forces:

(*a*) *Tectonic, or earth-building, forces* which are responsible for the main structural elements of the crust, *e.g.* mountain ranges, plateau blocks, and faulted depressions; such forces may be said to rough-hew the topography.

(*b*) *Erosional forces* which wear away, transport and deposit the material of the crust, and in so doing fashion the minor details of the surface relief; they may be said to etch, refine and polish the topography.

3. Denudation. The present landscape is only temporary, for even as tectonic forces raise a portion of the earth's crust, the surface comes under the attack of a variety of forces of destruction. Such processes eventually reduce all the elevations and fill up all the hollows of the earth's surface.

The general wearing away of the land surface by external forces or agencies is known as *denudation*, a term derived from the Latin *denudare* meaning "to lay bare." This general term embraces the results of *weathering*, *erosion* and *transportation*.

The general destruction of the earth's surface, which produces the plateaus and plains and the minor features or local land-forms, is carried out in various ways by the action of the sun, wind, rain, frost, water, gravity, etc. Three fairly distinct, but related, processes take place:

(a) *Weathering*, which may be defined as the weakening, breaking up, rotting and disintegration of the rocky material.

(b) *Erosion*, the scraping, scratching, grinding, battering and pulverising of the earth's surface rock.

(c) *Transportation*, the removal and dumping of loose surface material which is carried from one place to another by the agents of erosion.

WEATHERING

4. **Weathering and types of weathering.** Weathering is the loosening and breaking down of the rocky material of the earth's crust; it results in the formation of a covering of rock waste which is sometimes called *regolith*. Weathering occurs *in situ*, *i.e.* where the rocks are situated. Weathering, it should be emphasised, has nothing to do with the actual sculpturing and modelling of the land surface. That is the task of the agents of erosion.

Weathering occurs by two kinds of process:

(a) The disintegration of the rock by *mechanical* means;
(b) The decomposition of the rock by *chemical* means.

These two processes are seldom clearly separable; usually they work together, although one or the other may predominate under different climatic conditions.

Sometimes a third type of weathering is recognised: *biological* weathering. In this case both the mechanical and chemical aspects are involved.

Some of the more important ways by which weathering is effected are as follows:

5. **Temperature changes.** It is a law of physics that heat causes expansion. The alternation of extremes of temperature by day and night, such as occurs in hot desert regions helps to shatter rocks. The fierce sunshine heats the ground surface rapidly, for rock is not a good conductor of heat; hence the exposed rock

surfaces expand. Conversely, they contract when the sun goes down for the heat is quickly lost from the surface by radiation and so rapid chilling takes place. Furthermore, the mineral particles of which the rock is composed have different rates of expansion. Thus stresses and strains are set up in the rocks and, in due course, particles break off, flakes peel off (a process termed *exfoliation*), and sometimes blocks of rock split.

Mechanical weathering of this kind is most emphatic in arid regions and on exposed mountain tops: although laboratory experimentation leads us to believe that most mechanical weathering requires *some* moisture, however little, to be present to be really effective.

6. Frost action. When water freezes and turns into ice it expands in its volume; this simple physical change has tremendous consequences in weathering. Repeated freeze-thaw action has a weakening effect upon rock much like the temperature changes we have just noted. Water finds its way into the cracks and crevices and even the pores of rocks, and when it freezes it exerts a very strong pressure upon the rocks often causing them to split and shatter. Evidence of such frost action is to be seen in mountain areas, such as the English Lake District or the Alps, where the mountain slopes especially near the summits are frequently littered with angular rock fragments which have broken off from the rock face. Long-continued frost action on exposed mountain summits may result in the production of sharply pointed, shattered peaks known as *aiguilles* (needles).

7. Action of rain. The impact of heavy, torrential rain on soft rocks such as clay may exert some effect and sometimes small hollows made by raindrops can be seen in alluvium or clay. Where large pieces of rock or boulders occur in beds of clay or in loose sandy deposits, they may serve as protective "caps" with the result that *earth pillars* may be formed. In the boulder clay deposits forming cliffs, on the south side of Filey Brig, on the Yorkshire coast, there are numerous upstanding pillars to be seen, each capped by its protecting boulder. Sometimes the sheer physical impact of driving raindrops or hail may help to loosen particles. In general, however, the action of rain is probably more effective chemically than physically, although a certain amount of mechanical weathering would seem to be effected by falling rain.

8. Chemical weathering. Weathering includes many chemical processes, chief of which are the following: solution, oxidation, reduction, hydration, carbonation. The rate of chemical weathering varies greatly between different regions but is most pronounced in the hot, humid regions of low latitudes, for here the high temperatures and abundant moisture accelerate the speed at which chemical reactions take place. Chemical weathering would seem to be least effective in very cold and very hot regions, although *some* chemical weathering occurs everywhere on the earth's surface.

(a) *Solution.* Much chemical weathering results from the attack of weak acids. Rain, for instance, absorbs carbon dioxide as it falls through the air and so is converted into a very weak acid. But, weak though it may be, it is still strong enough to attack limestone rocks. The rainwater, now a weak carbonic acid, percolates through the joints in the limestone, which is soluble, and gradually dissolves it and eats it away. In this case carbonation has taken place.

(b) *Oxidation.* This is a common process of chemical weathering involving the combination of minerals with oxygen. The results of this action are most commonly to be seen when the rocks affected contain iron. The ferrous state changes into the oxidised ferric state to give an orange or reddish brown crust which crumbles easily.

(c) *Hydration.* This is the process by which minerals take up water and so expand, thereby causing stresses to occur within the rock which weakens its durability. The changing of olivine into serpentine is hydration.

(d) *Hydrolysis.* This is a process of chemical reaction involving water. In granite, for instance, the felspars disintegrate as a result of the chemical action of water and decompose into insoluble clay minerals, such as kaolin.

9. Biological weathering. Plants and animals may contribute to the weathering process: such biological action may be of either a physical or chemical nature. The prising action of tree roots and the trampling and borrowing action of animals help to weaken and break up the surface layers. Roots have remarkable penetrating and expanding power and not only are they often capable of prising the rock apart (the ruins of Angkor Vat in the Khmer Republic (Cambodia) show wonderful examples of trees and roots shifting huge blocks of masonry and splitting slabs) but they also provide openings which allow air and water to enter. The excretions of animals also provide acids which promote decay. The activities of bacteria and other minute organisms are especially significant in biological weathering. While, indeed, they are principally concerned with soil formation, locally they do assist in the decomposition and disintegration of the solid rock.

TABLE VII: FACTORS AFFECTING WEATHERING

1. *Geological Factors.*

    (a) *Parent rock.*   Rocks differ widely in their resistance to weathering due to both mechanical and chemical factors. Rocks made up of cemented particles are more likely to disintegrate under the weathering process than those composed of tightly interlocking crystals. Much also will depend upon mineral composition, *e.g.* rocks high in carbonates are susceptible to solution.

    (b) *Structure.*   Rocks of massive character are more likely to have a greater resistance to weathering than those which are bedded. Moreover, rocks characterised by fractures and fissures allow the percolation of water which undertakes solution.

2. *Topographic Factors.*

    (a) *Elevation.*   The higher the land the better chance is there for active water movement through the rock. In low-lying areas drainage may be poor; moreover, the water may become saturated with dissolved matter and so becomes incapable of effecting further solution.

    (b) *Slope.*   High, and especially steep, slopes promote downslope washing and the creep of weathered material. Rain-water tends to run off the surface rather than percolate downwards and so undertake some solution.

    (c) *Aspect.*   Slopes which are exposed to wind and rain are more prone to the action of weathering than sheltered slopes. Furthermore high altitude slopes facing the sun are more likely to suffer freeze-thaw action than poleward-facing slopes which are always sunless and cold.

3. *Mechanical Factors.*

(a) *Frost action.*   Water penetrates into rock pores, cracks and crevices and then freezes. Continual freeze-thaw action acts like a wedge forcing the rock apart. Freeze-thaw action also keeps the weathered material loose and porous; this, in turn, promotes the percolation of water which assists lubrication and helps the creeping, sliding and slumping of the unconsolidated material.

(b) *Temperature fluctuations.*   Rapid temperature changes, especially of extremes, may cause expansion and shrinkage of the constituent mineral particles of the rock or the surface face and so promote granular disintegration and exfoliation respectively.

(c) *Organisms.*   Plant roots may exert a wedging effect where they penetrate cracks in the rock. Burrowing animals may undermine and loosen soil and so make it easier for other factors to do their work.

4. *Chemical Factors.*

(a) *Temperature.*   High temperatures, especially if they are associated with plentiful moisture, promote chemical reactions and speed up rock decay. As a general statement it may be said that an increase of 10°C. (18°F.) roughly doubles the chemical reaction rate.

(b) *Rainfall.*   The higher the rainfall the more water will there be available to undertake solution and to react with the rock, *i.e.* to cause hydration, hydrolysis, carbonation, etc.

(c) *Organisms.*   Plants may act as a protective cover and so prevent the wasting of soil; on the other hand, their roots may assist the percolation of water and so help solution. Rotting vegetation may help to produce acids which attack the rock while the secretions of animals may also assist in a small way.

5. *Time Factor.*

Interval of exposure.   The length of time a rock surface has been exposed to the weathering process will affect the degree of weathering of a specific rock type and the thickness of the weathered mantle.

THE EROSION CYCLE

10. The agents of erosion. The various natural agents which carve and shape the earth's surface are:

(a) running water,
(b) groundwater and gravity,
(c) moving ice,
(d) wind action,
(e) waves and currents.

These agents not only undertake destructional work but also collect and transport the loose material produced by weathering and erosion. As they carry it away, the agents use the waste to erode the land surface; in other words, the agents use the material as "cutting tools" to carve up and shape the landscape.

11. The cycle of erosion. The sequence of events consequent upon the process of weathering, erosion and deposition, which result in the complete process of land change, from initial uplift of the land to its final destruction, is termed the *cycle of erosion*, or, less commonly, the *geomorphic cycle*.

The term cycle of erosion is a useful, all-embracing term which covers all the varied activities and transformations occurring upon the earth's surface as a result of modifying agents other than tectonic.

The American geomorphologist, W. M. Davis, enunciated the principle: "Landscape is a function of structure, process, and stage". By this he meant that the landforms which make up the landscape are the resultant of the interaction of the structure and character of the rocks, of the various denudative processes which modify these, and of the degree of transformation (*i.e.* the stage) to which they have been subjected.

12. The Davisian concept. The idea of the cycle of erosion, which was developed and formalised by Davis at the beginning of the present century, involved the transformation of the physical landscape through the natural agencies of weathering and erosion in an orderly progressive sequence of changes. In order to demonstrate how this cycle worked Davis introduced two assumptions:

(a) The cycle commenced on a newly uplifted land surface

and terminated when the land had been reduced to a level and almost featureless plain.

(b) The successive stages in this cycle passed through three phases of "youth," "maturity," and "old age."

13. Criticisms of the Davisian concept. Davis' ideas have been challenged in recent years and the main criticisms which have been levelled against his interpretation are:

(a) it is unlikely that a cycle can ever be completed because interferences, such as climatic changes or alterations in sea level, are bound to upset the orderly progress of the cycle;

(b) it is unlikely that land surfaces suffered rapid uplift and presented a smooth surface for the commencement of a cycle;

(c) it seems likely that the climate was at no time sufficiently stable over the prolonged periods required for a cycle to work itself out;

(d) no consideration was given to biogeographical influences (vegetation, soil and human activities);

(e) insufficient consideration was given to process and structure and too much emphasis was placed upon stages that lacked validity.

However, in spite of the various criticisms which have been advanced, it can be said that Davis' concept has provided a useful framework for the understanding of landscape development.

14. Polycyclic landscapes. Study of the land surface reveals that the idea of an erosion cycle does not correspond with the real conditions which regulate the activities of streams and are responsible for the fashioning of landscapes. Successive evolutionary stages (a progression from youth through maturity to old age) are probably exceptional rather than the rule. What in fact seems to have happened is that the evolutionary process has been slowed down, interrupted, brought to a standstill, and sometimes even reversed; this is shown by the occurrence of a series of river terraces which indicate that a series of cycles of erosion have taken place which have never been completed. Landscapes which show a series of uncompleted or partial cycles, overlapping in time, have been termed *polycyclic*.

15. Polygenetic landscapes.

As well as landscapes being polycyclic in character, many landscapes are polygenetic, *i.e.* they have had many origins or been subjected to several differing processes of erosion. As indicated above, it is doubtful whether there can ever have been any simple cycles since climate is unlikely to have remained sufficiently stable over periods long enough for a cycle to be completed. There are in many desert areas, for example, landscape features which clearly have resulted from water action; and one can only believe that these arid lands at one time suffered from a pluvial phase during which rainfall and run-off were responsible for much of the erosion and many of the landscape features. Again, in Britain, the landscape during the past 10,000 years or so has evolved under humid, temperate climatic conditions; but before that we know the land was greatly affected by ice during the period of the Pleistocene Glaciation. Hence, the present landscape is very largely the outcome of the work of two very different processes of erosion: one by glacial action, the other by river erosion. Thus regions whose geomorphological features have undergone several differing processes of erosion are described as *polygenetic*.

PROGRESS TEST 11

1. "The everlasting hills." Comment upon the accuracy of this statement. (1)

2. What are the two fundamental physical forces responsible for landscape development and change ? (2, 3)

3. What are the three processes involved in denudation ? (3)

4. What is meant by weathering and how does it take place ? (4)

5. Describe how temperature changes influence weathering. (5, 6)

6. "Most mechanical weathering requires some moisture, however little, to be present to be really effective." Briefly discuss this statement. (5)

7. Enumerate the chief chemical processes involved in chemical weathering and explain briefly how these processes work. (8)

8. Write a brief account of the action and significance of biological weathering. (4, 9)

9. What is meant by the phrase "cycle of erosion" ? (11)

10. What is meant by the expression: "Landscape is a function of structure, process, and stage" ? (11)

11. Outline the criticisms which have been levelled against W. M. Davis' concept of the erosion cycle. (12, 13)

12. Briefly explain the meaning of polycyclic and poylgenetic landscapes. (14, 15)

CHAPTER XII

GROUNDWATER, GRAVITY AND MASS MOVEMENTS

GROUNDWATER

1. The hydrological cycle. The hydrological cycle (*see* Fig. 38) connotes the movement of moisture from the ocean surfaces (or any other water surface) to the land, where it is precipitated, and its eventual return by evaporation and run-off to the sea.

Water is evaporated by the sun and the wind from the ocean surfaces, rises by convection and is carried by air movements towards the land. Over the land the water vapour in the air masses undergoes cooling and condensation, either by convection or orographic or frontal lifting. Ultimately the atmospheric moisture is precipitated as rain, hail, snow, etc. This surface moisture does one of three things:

(*a*) It is re-evaporated before it has time to sink in or run off the land surface.

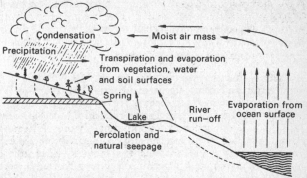

Fig. 38.—*The hydrological cycle* involves the movement of moisture from the ocean surface to the land where, by means of evaporation, percolation and run-off, the water is, sooner or later, returned to the ocean. The whole system is of considerable complexity.

(b) It sinks into the soil and percolates downwards into the rocks beneath the surface.

(c) It runs off the land surface in the form of sheet wash, rills and streams.

At the same time, on land, transpiration from plants and evaporation from the soil and water surfaces such as lakes, ponds and rivers adds water vapour to the air.

The moisture that goes underground, through percolation and natural seepage, usually breaks through to the surface again, e.g. in the form of springs, lower down the slope of the land surface; this spring water feeds streams which eventually reach the sea.

2. Underground water. Water, in lesser or greater quantities, is present everywhere in the soil, subsoil and bedrock: such water is known as *groundwater*. It is of either *external origin* (derived from atmospheric or surface waters) or *internal origin* (derived from the interior of the earth).

Substantially the greatest amount of groundwater comes from atmospheric precipitation; this is called *meteoric water*. The water retained in sedimentary rocks from the time when they they were originally laid down is termed *connate water*. Water liberated as a result of igneous activity, e.g. during the crystallisation of magma, is called *juvenile water*.

The vast supplies of artesian water which occur in some places may be residual water left behind when sediments were deposited which has remained locked up in the rocks but more commonly it is the result of meteoric water which has percolated underground through porous or permeable rocks.

3. Factors affecting percolation. As already mentioned, of the rainwater and other forms of precipitation reaching the ground some is evaporated back into the atmosphere, some percolates into the ground, and some runs off into the streams. In Britain, the proportions are approximately one-third each; but in other parts of the world the proportions may vary widely.

The amount of water entering the ground varies greatly from place to place and is determined by the following factors:

(a) The amount and nature of the rainfall.

(b) The slope of the land surface.

(c) The porosity and permeability of the surface layers.

(d) The rate of evaporation.

(e) The amount and nature of the vegetation cover.

(f) The amount of water already in the soil.

4. The movement of groundwater. Water seeps slowly downwards through the pores in the rocks and through fissures until it is halted by a layer of impervious rocks (*see* Fig. 39). A water-holding bed of rock, that is, a bed of rock which holds water

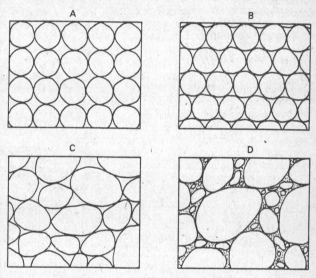

FIG. 39.—*Porosity in rocks.* (a) and (b) show the effect of the packing of grains upon pore-space; the interstices between the grains in (b) are much smaller than in (a). (c) illustrates a natural sand with a good sorting of grains, thereby giving a high degree of porosity while (d) shows a poorly sorted sand with a matrix of clay giving a low degree of porosity.

after the fashion of a sponge, is called an *aquifer*. Three zones above an impermeable bed are usually distinguished:

(a) The upper zone through which the water descends and which has air-filled pores; this is called the *zone of aeration* or the *vadose-water* (*i.e.* wandering water) *zone*.

(*b*) The lower zone where water accumulates and from which the air has been expelled; this is termed the *zone of saturation*.

(*c*) The in-between zone, which contains groundwater after long-continued rain but dries out during periods of drought; this is called the *zone of intermittent saturation* (*see* Fig. 40).

The upper surface of the zone of saturation is called the *water-table*. Its height varies from place to place and from season to season. The water-table reflects, in a subdued way, the land surface profile, rising beneath the hills and dropping under the valleys. The water also moves horizontally from under the hills, where the water-table is highest, towards the valleys. This lateral movement is very slow since the water moves usually by percolation through the pores in the rock. These processes of downward and lateral movement are illustrated in Fig. 40.

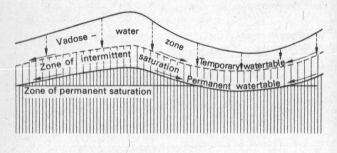

----▶ Downward percolation of water

——▶ Horizontal movement of water

Fig. 40.—*The movement of groundwater.* Water percolates both downwards and horizontally. The permanent water-table marks the level of permanent saturation, but in periods of abundant rainfall the water-table will rise to give a temporary water-table. Note how the water-table reflects, in a subdued way, the profile of the land surface.

5. Phenomena associated with groundwater. Groundwater, in finding its way underground, gives rise, especially in soluble rocks such as limestone, to surface phenomena such as swallow-holes, potholes and poljes; these, in turn, often give rise to subsidence depressions and collapsed caverns (*see* Fig. 41).

The erosive activity (chiefly solution) of water as it works its way through the rock strata sometimes produces "underground scenery"; this is most spectacularly developed in limestone regions, especially in Carboniferous and Oolitic Limestone, producing subterranean passages, large chambers sometimes with pools and waterfalls, stalagmites and stalactites.

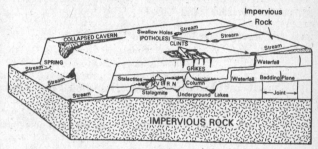

Fig. 41.—*Features found in a limestone area.* The diagram shows some of the characteristic features (caves, potholes, underground streams, disappearing streams, Vauclusian springs, etc.) found in a limestone or karst area such as Malham in Yorkshire, the Causses region in southern France, or in Dalmatia in Yugoslavia.

Stalactites are icicle-like formations composed of calcite which hang vertically from the roofs of caverns and passages. They are formed by the constant dripping of water, charged with calcium bicarbonate, which has seeped through the rock above. The stumpy, roughly coneshaped mass formed on the cave floor by the dripping from the stalactite above is termed a *stalagmite*. Growth of both forms is very slow—taking thousands or tens of thousands of years—but sometimes the two may join to form a pillar or column. Sometimes these and other deposits of calcite are beautifully coloured as a result of the staining effects of iron, copper, etc.

GRAVITY AND MASS WASTING

6. Gravity. Everywhere on the earth's surface gravity exerts a downward pull on all materials. Solid and loose mantle materials are most easily affected, but sometimes even the solid bedrock is affected. The term *mass movement* or *wasting* is given to all the different kinds of downslope movement due to the pull of gravity. The tendency to mass wasting is related to a number of conditions:

(*a*) When the soil and the mantle are loose or poorly consolidated and deeply weathered.

(*b*) The angle of repose of the regolith (loose, partially broken mantle rock) and the degree of inclination of the strata.

(*c*) Inter-banded hard and soft rocks, especially when clay (which can be greasy when moist) is in the sequence.

(*d*) Thin beds increase the tendency to movement because there are more bedding-planes along which slipping can occur.

(*e*) The amount of moisture present: movement is greatest in very wet weather, especially after long, dry spells.

(*f*) The absence of a substantial cover of vegetation, for plant roots help to hold the soil in place and give it a measure of stability.

A number of other factors also assist downslope movement: the heating and cooling of the soil, the freezing of the soil and subsoil, frost heaving (*i.e.* the "lifting" action of frost), the tramping and burrowing of animals, the shaking produced by earth tremors and even the clap of thunder.

7. Types of mass movement. Six main kinds of mass movement are recognised: creeping, flowing, sliding, slumping, rock falls and subsidence.

(*a*) *Soil creep.* Slow, downhill movement of the soil and weathered mantle occurs on any moderately steep slope. It is an imperceptible but continual movement. *Terracettes* found on grassy slopes (erroneously attributed to the action of sheep) are formed by soil creep. The creep of loose rock fragments produces *talus* or *scree*.

(*b*) *Earth and mud flow.* Earth flow or *solifluction* is the downslope shift of large masses of water-saturated soil during a period of a few hours. Mud flows are streams of semi-liquid mud which often flow down valleys in mountain areas.

(*c*) *Landslides.* The rapid sliding or slipping of large masses of rock or mantle is termed a landslide or landslip. It is a

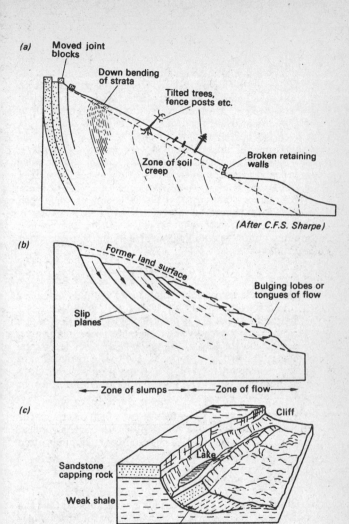

(a) Moved joint blocks

Down bending of strata

Tilted trees, fence posts etc.

Broken retaining walls

Zone of soil creep

(After C.F.S. Sharpe)

(b) Former land surface

Slip planes

Bulging lobes or tongues of flow

← Zone of slumps → ← Zone of flow →

(c) Cliff

Lake

Sandstone capping rock

Weak shale

Plane of slip   (After Strahler)

FIG. 42.—*Mass movements.* (a) illustrates the effect of soil creep. (b) shows how slumping may take place along curving slip planes producing bulging lobes of earth. (c) illustrates how a massive block of land may slump downwards from a cliff face.

movement characteristic of steep slopes. Landslips are especially common along stream banks and escarpments.

(d) *Slumping*. Closely related to landslips are *slump blocks*, massive blocks of bedrock which break off from cliffs and, as they slide down, rotate or tilt backwards along a plane of slip; the shearing surface is often spoon-shaped.

(e) *Rock falls*. The free fall of masses of rock or individual boulders from cliff faces or down steep hill slopes. Such rock masses, loosened by weathering, usually shatter into smaller fragments and accumulate as talus at the base of the cliff or slope.

(f) *Subsidence*. The vertical displacement of rock with very little or no horizontal movement is termed subsidence. It is usually a slow, settling movement resulting from mining or draining, but occasionally may be rapid, *e.g.* the collapse of underground caverns.

SLOPES AND THEIR SIGNIFICANCE

8. Slopes. While some slopes may be of tectonic origin or result from erosion, many are associated with gravity and its effects. In recent times the geomorphologist has directed his attention to slopes, especially their characteristics and origins, for he sees in them the key to the understanding of landscape development. This study of slopes has led to the recognition of four possible elements in a hillside slope profile. Figure 43 illustrates these elements in an idealised form; from hill top to valley floor they are as follows:

(a) The *waxing slope* or convex slope which occurs at the top of a hill where the slope curves over to meet the vertical face below. This uppermost slope is soil-covered and is often spoken of as the *upper wash slope*.

(b) The vertical, wall-like exposure of bare rock immediately below the waxing slope is referred to as the *free face*. The slope here is too steep to allow any accumulation of weathered material such as scree. This free face is sometimes termed the *derivation slope*, since any rock waste littering the lower slopes comes from this face.

(c) The *constant slope* is the straight slope of the lower hillside which lies at the angle of rest of weathered material. It is usually called the *debris slope*, since this is where scree and other weathered debris tends to accumulate.

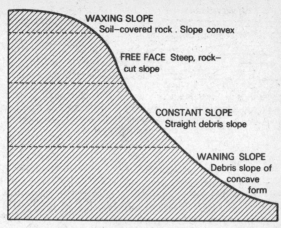

WAXING SLOPE
Soil—covered rock . Slope convex

FREE FACE Steep, rock—
cut slope

CONSTANT SLOPE
Straight debris slope

WANING SLOPE
Debris slope of
concave
form

FIG. 43.—*Elements of slope*. The figure shows, diagrammatically, the
four facets of slopes: the waxing slope, the free face, the constant slope,
and the waning slope.

(*d*) The *waning slope* at the foot of the hill is a gentle slope
of concave form resulting from the washing down and
accumulation of, usually, the finer debris. The slope merges
into the valley floor. It is frequently referred to as the
lower wash slope.

9. The significance of slopes. Reference was made in Chapter
XI to W. M. Davis's concept of landscape development, a
theoretical interpretation of the formation of landforms, more
particularly in cool, moist environments, which largely dominated ideas in North America and Britain until the 1950s.
Landforms in humid temperate regions tend to show rounded,
subdued outlines, the outcome of, according to Davis, "down-
wearing," *i.e.* downward erosion.

Some geomorphologists, notably German workers such as
Penck, who studied the shapes of landforms in the tropical
regions, were impressed by the flat tops, flat floors, steep slopes
and sharp angles of the topography and arrived at the conclusion
that the landforms were largely the result of "back-wearing,"
i.e. lateral erosion (*see* Fig. 44). This interpretation was completely at variance with Davis's ideas. Here we cannot go into

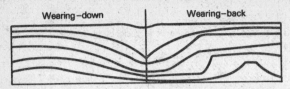

Fig. 44.—*Down-wearing and back-wearing.* The diagram illustrates the contrasting cross sections of valleys from "youth" to "old-age": on the left according to the process of wearing-down and on the right by the process of wearing-back.

the whole complex problem of landscape development, but the crux of the argument is as to the manner by which slopes have been formed and the relief has been lowered: whether, according to Davis, by a process of essentially progressive flattening or, according to Penck, by essentially parallel planing. It is by studying slopes that the geomorphologist seeks to arrive at the truth of landscape development.

PROGRESS TEST 12

1. Explain what is meant by the expression "the hydrological cycle". (1)

2. Describe what may happen to precipitation when it falls on the land surface. (1)

3. Name and account for the origin of the different kinds of ground-water. (2)

4. Cite the factors which are responsible for the amount of water entering the ground at a given place. (3)

5. Describe and explain the different movements of water underground. (4)

6. What is meant by the expression "underground scenery"? Describe some of the features of underground scenery. (5)

7. What conditions predispose towards the different kinds of down-slope movement? (6)

8. Summarise the six principal types of mass movement. (7)

9. Show how gravity and groundwater may be responsible for down-slope movements. (6, 7)

10. Describe, with the aid of a diagram, the different elements which are recognised in a hillside slope profile. (8)

11. Explain the following terms: water-table, stalactite, connate water, solifluction, back-wearing. (2, 4, 5, 7, 9)

12. What is the importance and significance of the study of slopes? (9)

FLUVIAL PROCESSES AND LANDFORMS

STREAM EROSION

1. The physical functions of rivers. Rivers and their associated streams undertake two very important physical functions:

(*a*) By draining the land surface they dispose of the superfluous water brought by precipitation.

(*b*) They are responsible for much of the denudation of the land surface over large parts of the earth.

(*i*) They dissolve and erode the rocks over which they flow.

(*ii*) They transport the matter which they have dissolved or eroded away.

(*iii*) They deposit some of the material which they carry in suspension or roll along the stream bed.

2. Rivers and valleys. In 1802 John Playfair laid down the simple law that where two rivers join each other (their confluence) they do so at a common level; in other words, at the point of confluence the upper level of water in one river is precisely the same as the upper level of water in the other. Such a statement may appear to be stating the obvious, to be trivial and quite unnecessary. However, Playfair's law is important for it infers that all valleys have been fashioned by the rivers which flow in them or, expressing this in a slightly different way, rivers are the causative agents of the formation of their valleys. If valleys were not the result of the rivers flowing in them, then it would be possible for valleys to meet one another at any relative level, but this is never so except under exceptional circumstances, *e.g.* where glaciation has resulted in "hanging" valleys. Thus Playfair's simple law recognises the cause-and-effect relationship between rivers and their valleys.

3. The erosive work of streams. The discharge, the velocity and the load (*i.e.* the material carried or moved) of a stream are all significant in the work of erosion.

Streams derive their supplies of water, directly or indirectly, from rainfall and other forms of precipitation, *e.g.* melting snow and ice. It is well known that, at least in humid environments, streams increase their volume as they proceed to the sea but that this volume is likely to vary according to the season of the year. From an erosional point of view, it is the velocity of the stream or the speed of movement of the water that is particularly significant. Swiftly flowing water erodes more effectively than gently flowing water.

The volume of water carried and the speed of movement determine the energy of a stream and upon this energy depends the carrying capacity of the stream and the nature of its load.

4. Stream load. The amount of material carried or moved by running water is of very great importance, for upon this very largely depends its capacity to undertake erosive work. It should be understood that running water in itself, apart from cavitation (*see* 5 (*b*) below) and any solution that it may effect, has little erosive action, although it may undertake a certain amount of hydraulic activity. Once, however, water has become charged with debris (which acts as its cutting tools), it begins to wear away the surface over which it is running.

Generally speaking, it may be said that the erosive capacity of running water increases in proportion to the following:

(*a*) Its velocity, *i.e.* its rate or speed of flow.
(*b*) Its load, *i.e.* the amount of matter carried.

NOTE:
(*i*) We must be wary of accepting (*b*) as it stands, since a river carrying too great a load leads to *aggradation* (building up) and the protection of the channel bed.
(*ii*) A stream is able to carry a much greater quantity by weight of fine material than of coarse. The amount of solid matter which a stream is capable of moving is termed its *capacity*. The term *competence* is used to refer to the maximum size of particles capable of being moved by the stream.
(*iii*) The greater the stream's energy or power, the greater is its capability to move coarse material. The lowest velocity at which rock particles of a given size move is termed the *critical erosion velocity*.

5. The mechanism of stream erosion. The erosional work of running water acts in several different ways:

(a) By *solution*, *i.e.* the solvent action of water as it flows over the rock. The degree of solution varies greatly, depending upon:

(i) The purity of the water and (ii) the solubility of the rock.

(b) By *cavitation*, *i.e.* the collapse of air bubbles in turbulent water; the bursting of such bubbles produces "shock waves" which help to weaken and disintegrate adjacent rocks.

(c) By *hydraulic action*, *i.e.* the lifting or quarrying effect of rushing water. The sheer force of moving water may dislodge rocks.

(d) By *corrasion* or the abrasive action of waterborne matter, *i.e.* material carried in suspension or rolled along the river bed.

(e) By *impaction* or the effect of blows upon the river bed or banks by large boulders bounced along during floods.

(f) By *attrition* or the shattering and breaking-up of the stream load through collisions and mutual abrasion.

6. Stream transportation. The load carried by running water is moved downstream in a variety of ways (Fig. 45):

(a) In solution.
(b) In suspension.
(c) By *saltation*, *i.e.* movement by a series of jumps.
(d) By *traction*, *i.e.* by being rolled along the river bed.

Paths of particles moved by saltation

Suspended load

Rolling of pebbles – transport by traction

Bulk movement of sand bed load as small dunes – traction

After Rogers and Adams

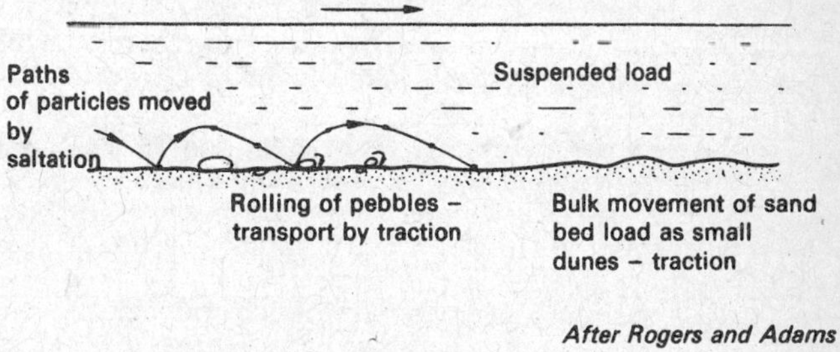

FIG. 45.—*Stream transportation.* Stream material is moved downstream in four different ways: (i) some material (apart from that which is dissolved, *i.e.* carried in solution) is carried in suspension; (ii) some is moved in a series of jumps *i.e.* by saltation; (iii) some is rolled along the bed by traction; and (iv) some is moved by traction in bulk.

As a general rule, however, the amount moved by traction and saltation is much less than that transported in suspension and solution. The following figures which relate to the Mississippi are interesting and indicate the enormous amount of material which may be carried by a great river, although we should be wary of accepting these as being typical of a large river; the average annual discharge is estimated at 340 million tons of matter in suspension, 156 million tons of matter in solution and 40 million tons of matter by saltation and traction. The total amount of sediment carried to the oceans by all the rivers of the earth cannot be precisely worked out but a rough estimate has been made and this amounts to the order of 10^{10} tons.

7. River deposition. Whenever flowing water is reduced in speed some of the load is deposited but it is also clear that a decrease in turbulence (the whirls and eddies which occur in running water) leads to deposition too. The heavier, coarser material is the first to be jettisoned, the fine silts and muds last. It should be remembered that material can be, and is, dropped at any point along a river's course: deposition is not confined to the lower course where, however, it is most usually evident. Deposition in the upper courses of streams is purely temporary and, sooner or later, will be removed.

Checks in the velocity of stream flow, leading to the deposition of material, may occur at the following points in a river's course:

(a) At a break in slope, *e.g.* where a stream leaves the hills and enters a plain: at such points alluvial fans are often built up.

(b) Where the valley floor widens and allows flooding to occur which results in alluvium being deposited.

(c) Where a swiftly flowing stream enters the quiet waters of a lake; this leads to the sudden dropping of sediment (seen in lake deltas).

(d) Where streams peter out on entering arid regions, thereby causing the fluviatile debris to be jettisoned.

(e) Where rivers debouch into the sea, the salt in the water helps clay particles to flocculate or coagulate to form larger grains which are likely to settle more rapidly and so help to build up deltas.

VALLEYS AND VALLEY FEATURES

8. The character of valleys in humid landscapes. The American geographer, W. M. Davis, studying landscapes in moist, temperate regions, recognised three stages through which a stream passes and three associated valley profiles and cross-sections (Fig. 46). He likened these stages to the stages in a

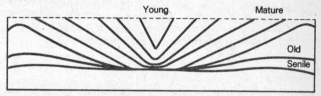

Fig. 46.—*River valley cross-section.* Cross-section to show the stages in the development of a valley. As the valley deepens and widens so the interfluves are correspondingly reduced and flattened out.

life-cycle: the stage of *youth* when a river is in its upper or torrent course, the stage of *maturity* when a river is in the middle valley course; the stage of *old age* when a river is in the lower or plains course. This concept is a useful one but we should remember that it is largely an idealised cycle and an idealised division, because no two rivers are alike. The lengths and features of different sections of rivers (and their associated valleys) vary considerably.

(*a*) The *torrent* or *upper course* occurs where the volume of water is small, the stream often swiftly flowing, the gradient is steep (roughly about 25 ft. (7·62 m) or more per mile, and the valley cross-section is V-shaped. In this section active headward erosion usually occurs, *i.e.* the stream valley is lengthened by the stream eating back in an up-slope direction. Here, also, in the torrent section vertical, *i.e.* downward, erosion is more important than lateral or sideways erosion. Cascades, waterfalls and rapids are characteristic features of the streams, which wind between interlocking spurs.

(*b*) The *middle* or *valley course* occurs where the volume of the stream is much greater (because of the increase in the catchment area), the gradient has lessened (drop is of the order of 10 ft (3·04 m) per mile), lateral cutting becomes active, and the stream begins to wander in curves or meanders on the floodplain it has cut out. In this section the interlocking spurs suffer destruction as the stream meanders move down valley and some deposition occurs. As the river continues to erode downwards towards its base-level, valley widening occurs due to:

(*i*) the lateral cutting action of the river and
(*ii*) by rain-wash and rills which carry loose weathered debris downslope.

(*c*) The *lower* or *plains course* occurs where the volume of water is great, the river is wide, the gradient imperceptible (often less than 1 ft (0·3 m) per mile) and where downward cutting has ceased and deposition becomes the main work of the river. Aggradation or the dropping and accumulation of the river's load takes place. Characteristic features of the river floodplain are meander scars or scrolls, ox-bow lakes, and levees (raised embankments).

9. The character of valleys in arid landscapes. The same stages in the life cycle of a river occur in semi-arid and arid areas but the valley forms tend to be different. In dry regions it would seem that the widening of valleys is almost entirely the result of direct stream erosion for there is little of the downslope movement of material such as is typical of humid areas. The mass movement of material downslope is less because:

(*a*) there is less loose, weathered material available for movement (wind carriers much of the finer matter away);
(*b*) there is less moisture in the soil and mantle to assist lubrication and downslope movement;
(*c*) there is little development in arid regions of soil.

Hence, in dry regions, the characteristic river valley has sides which are quite steep irrespective of the stage of development of the river itself.

Whereas in humid environments the lower hill slope merges, usually, into the valley floor by means of a gentle concave slope, in arid regions the break between hill slope and valley floor is sharp or angular. Often a *pediment* is formed (Fig. 47); this is a flat or very gently sloping surface at the foot of the valley wall of bare rock and this appears to have been formed by the abrasional action of sheets of floodwater; in other words, the sheets of water

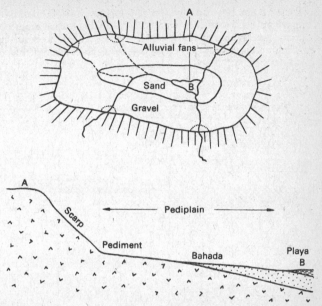

FIG. 47.—*Pediment.* The upper diagram shows an intermontane basin in an arid area; streams, often of an intermittent nature, carry rock debris into the depressions and build up alluvial fans around the periphery. The lower diagram gives a cross-section from the plateau to the salt-pan in the depression.

wash debris over the surface and scour the rock into a flat pediment.

Also characteristic of valleys in dry areas are *gravel* and *alluvial cones* and *fans*. Because of the dry climate, stream activity in arid regions is very irregular. The intermittent streams, however, bring down and dump large quantities of debris at the base of the hill slopes bordering the valley, building up triangular-shaped deposits.

10. The long profile. The long profile of a river, *i.e.* from its source to sea, takes the form of a smooth curve in the case of mature rivers (*see* Fig. 48). Such a profile is known as a graded profile or *thalweg* and implies that a condition of balance or equilibrium has been achieved, *i.e.* the river is neither eroding or depositing. Clearly, however, there is always *some* erosion and *some* deposition taking place but if one views this over a long term such erosion and deposition cancel each other out and so

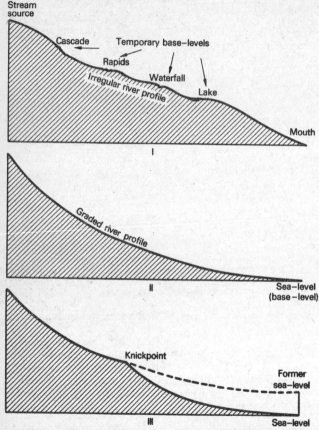

FIG. 48.—*Long profile of a river.* I shows the irregular long profile broken by rapids, waterfalls and lakes. II shows the graded profile when all the obstructions have been eliminated. III shows how rejuvenation has led to a new base level being formed.

in this way a balance or equilibrium is achieved. This concept of a graded condition is an indication that river maturity has been reached. The following conditions are indicative of maturity:

(a) The absence of waterfalls, rapids and lakes (ox-bows excepted).
(b) The occurrence of a floodplain with natural levees.
(c) The presence of meanders, meander scrolls and ox-bow lakes.
(d) The valley sides are subdued or gently sloping.
(e) The maintenance of stable channel characteristics.

11. Rejuvenation. Since a river is constantly grading its bed it would, theoretically, finally reach its base level of erosion: in other words, the river would be flowing at or very near sea level. This is rarely, if ever, attained for the balance may be upset by climatic changes or tectonic uplift. Rejuvenation of a river is most commonly caused by either the elevation of the land or the sinking of the sea level. Any uplift, for instance, would result in the steepening of a river's gradient which would cause an increase in stream velocity and this increased energy, in turn would be spent in renewed erosion.

Rejuvenation results in the following:

(a) *Incised meanders*, that is, stream meanders entrenched between steep, and usually symmetrical, sides.
(b) *River terraces*, that is, steps on either side of the valley marking former floodplains.
(c) A *valley cut within a valley*—the first stage in river terrace development.
(d) A slight *break in slope* and *change in the shape of the valley*.

FLUVIATILE LANDFORMS

12. Waterfalls. A waterfall is a steep fall of river water marking a sudden, and often spectacular, interruption in the river's course. Falls occur wherever a stream's efforts to achieve a graded profile are interrupted. Falls are essentially temporary features of the landscape. Waterfalls may result from a variety of causes:

(a) By the differential resistance of rocks to erosion as when:

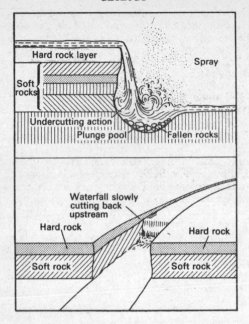

FIG. 49.—*Waterfalls*. The upper figure shows the undercutting action of falling water; the softer layers are eroded away more quickly than the more resistant capping rock. The lower figure shows how a waterfall slowly cuts its way upstream.

  (*i*) there is a resistant cap rock, *e.g.* Niagara.
  (*ii*) a bar of resistant rock, such as a dyke, cuts across the stream channel, *e.g.* Cauldron Snout, in Teesdale.

 (*b*) By the occurrence of faulting, *e.g.* Gordale Scar at Malham, the Zambezi Falls.
 (*c*) By glaciation which has led to the formation of hanging valleys tributary to the main valley, *e.g.* the falls of the Lauterbrunnen valley in Switzerland.

In due time all waterfalls cut back up valley and become eliminated to form part of the graded valley.

13. River terraces. Terraces are common features of river valleys. They result from rejuvenation which causes rivers to carve out new floodplains within their already existing floodplains. Each time a phase of rejuvenation takes place, downward cutting occurs and the creation of a new floodplain commences. A river valley may show a whole sequence of terraces which have been developed in this way.

Terraces are particularly well developed along the River Rhine and can be traced in the lower Thames valley around London where three distinctive terraces can be distinguished— the Boyn Hill Terrace, the Taplow Terrace, and the Floodplain Terrace. The highest terrace is always the oldest, *i.e.* the first to have been formed. The height of one terrace above another varies: sometimes the difference may be a matter of only a few feet, at other times it may be 50 ft (15·2 m) or more above its neighbour. Frequently it is possible to correlate the terraces along the sides of a valley with the *knickpoints* (breaks of slope) in the long profile.

14. Meanders and ox-bows. On many river floodplains, such as the Thames, Seine, and Mississippi, the rivers show winding or looping courses: such curves are termed meanders. Such meanders may be of any size though some are of great size, *e.g.* 5 miles (8 km) in diameter. Meanders have certain recognised geometrical attributes, *e.g.* the radius of curvature is equal to twice or three times the stream width, the ratio between the wave length of the meander and the radius of curvature is constant. What, precisely, causes rivers to meander has long been debated, but it would seem that they are associated with *pools* (stretches of quiet water) and *riffles* (stretches of disturbed water) and rivers which carry a high proportion of their load in suspension. Meander channels show steep outside banks since the current impinges most strongly on the concave side; this leads to maximum erosion at this point and undercutting may occur to cause the formation of a bluff or river-cliff. Conversely, on the inner side of the curve, where there is slack water, deposition may occur. This more gently-sloping inner side of the bend is termed the slip-off slope.

Since a river's current is always directed towards the outer, opposing banks of adjacent meanders, the curves gradually approach each other until only a narrow *swan's neck* divides them. Ultimately, during flood time, the water cuts across these necks and carves out a fresh channel, thus shortening the river's course. The former bend is now called an *abandoned meander* and it becomes separated altogether from the flowing water by deposited alluvium to form an ox-bow lake or *mortlake* (dead lake). Examples of ox-bows are very common on the lower Mississippi floodplain which is also scarred with the remains of older ox-bows which have dried up.

15. Levees. These are natural embankments which are built up by rivers along the edges of their channels. They are caused

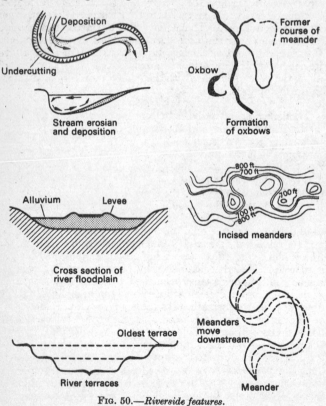

Fig. 50.—*Riverside features.*

by river floods. When a river reaches its bankfull stage and the
water spills over on to its floodplain, alluvium is deposited
along its banks and gradually builds up to form a kind of
containing wall. When the flood waters subside, these naturally
built banks or levees remain. With each successive flooding
they grow in height. They have the effect however, of causing
the bed of the river to rise above the level of the adjacent or
surrounding floodplain. Hence if the levees are breached during
subsequent times of high water, the water will spill over on to
the floodplain causing flooding. Moreover, this flood water
cannot drain back into the river because it is prevented from
doing so by the barrier created by the levee. In the past
extensive and serious flooding has occurred along the Hwang-ho
and Mississippi when the levees have been breached.

NOTE: the term levee is also applied to artificial, man-made embankments which sometimes have been constructed along the banks of
rivers to check their flooding.

16. Deltas. A feature found at the mouths of many rivers
(and also in lakes) is a delta. There are many misconceptions
about delta formation, chief of which are that they only occur in
tideless seas and that long plains courses and sluggish flow are
necessary. These ideas are untenable. Two main conditions are
necessary for river-mouth deposits: there must be an appreciable
load of silt being brought downstream and the silt must be
allowed to accumulate on the gently shelving sea floor.

A number of factors or conditions do dispose towards the
deposition of deltaic sediments:

(a) The quantity of sediment supplied.
(b) The depth of the water at the river's mouth.
(c) The strength of waves and currents.
(d) The extent of the tidal range and its influence.
(e) The flocculation (the process by which particles of
     sedimentary matter coagulate) caused by salt water.

These factors help to control the shape, areal extent, and volume
of the deltaic deposits.

17. Types of deltas. If the conditions are favourable a delta
will be built up consisting of a threefold succession of deposits
termed bottomset, foreset and topset beds. A delta may be

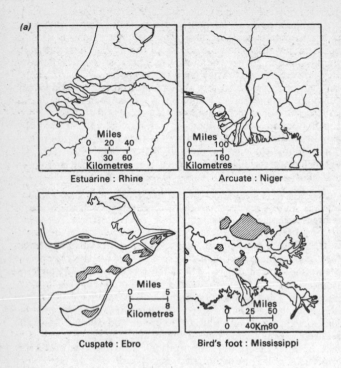

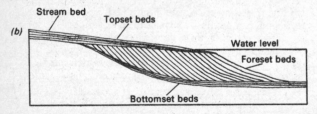

FIG. 51.—*Types of delta.* (*a*) here are four different shapes of delta; precisely what shape a delta takes is mainly due to the conditions of water circulation in the area where the river debouches into the sea. (*b*) a cross-section of a delta (schematic).

defined as a level area of alluvium which has accumulated at the
mouth of a river where the deposition of some of its load exceeds
its rate of removal, crossed by distributaries—the divergent
channels of the river. Four types of delta (*see* Fig. 51) are
sometimes distinguished:

(a) *Estuarine*, *e.g.* Rhine, Amazon where infilling of an
estuary occurs.
(b) *Arcuate* or fan-shaped, *e.g.* Nile, Niger.
(c) *Cuspate* or pointed like a tooth, *e.g.* Ebro.
(d) *Bird's-foot* type with fingering branches, *e.g.* Mississippi.

STRUCTURE AND DRAINAGE

18. Rivers and structure. Rock structure exerts an important
influence upon the activities and patterns of rivers. As the
underlying beds of rock of varying resistance become exposed
as a result of subaerial activity, the land surface becomes
diversified into upstanding ridges and hollows to produce for
instance, the scarp and vale topography of lowland England.
As streams flow over the land surface they adjust themselves
to the rock structure; for example, where they cross ridges their
valleys will be relatively narrow and deep, but where they cross
the depressions the valleys commonly will be wide and shallow.
Banded rock outcrops of this type give rise to a rectilinear or
trellis drainage pattern as may be seen in the Weald. Where the
rock is of a homogeneous nature, a branching or dendritic
pattern emerges. In the case of dome structures or on volcanoes
the streams flow outwards radially from a central point to
produce a radial drainage pattern as may be seen in the Lake
District, which is a dissected dome, or on Dartmoor.

The process of river capture, alternatively termed piracy or
beheading, is a common feature of many river systems; this
occurs when a river acquires the headstreams of a neighbouring
river, thereby enlarging its own drainage basin at its neighbour's
expense. This may happen through headward erosion (or
watershed migration) or by a stream tapping and diverting the
waters of a neighbouring river.

19. Superimposed and antecedent drainage. Rivers normally
adjust themselves to the surface structure of the land they drain
but sometimes they appear to show no respect for the surface

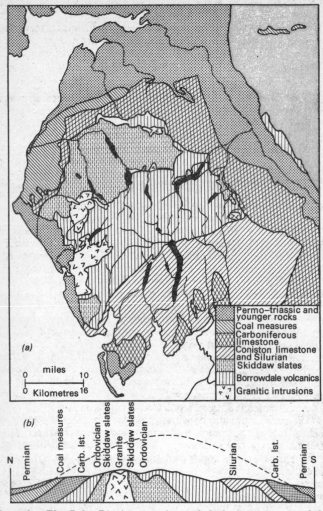

FIG. 52.—*The Lake District; superimposed drainage.* (a) an eroded dome with radial drainage; the present drainage pattern originated on the Carboniferous Limestone strata which covered the older, harder rocks beneath. (b) a geological cross-section of the area.

structure and follow courses quite unrelated to the strata they cross. For example, a river may traverse a resistant outcrop of hard rock several times or erode a valley for no apparent structural reason. When a river behaves in such a way, ignoring the structure, it usually implies that it must have originated upon an overlying rock surface which no longer exists. Where the influence of the original rock structure continues to predominate over that of the newly exposed strata, superimposition is said to have occurred and the drainage is described as superimposed drainage. The drainage of the Lake District is a case in point, (*see* Fig. 52).

Rivers sometimes seem to show complete independence of relief, as when a river, such as the Indus, cuts through a high mountain range. It is believed that in such cases the river was there *before* the mountains were built; in other words, the river's erosive action kept pace with the rate of uplift. Such rivers, which were established before the relief features were developed and are therefore discordant with the present structure, are said to have antecedent drainage.

PROGRESS TEST 13

1. What are the physical functions of rivers? Note the three kinds of work which rivers perform. (1)

2. State Playfair's law and say why it is important. (2)

3. Briefly describe the various ways in which rivers undertake erosive work. (5)

4. Describe the meaning of the following terms: stream capacity, stream competence, flocculation, aggradation, attrition. (4, 5)

5. In what different ways is stream load carried or moved? (6)

6. Checks in the velocity of stream flow, leading to deposition, may occur at a number of points along a river's course. Briefly describe these various check points. (7)

7. Briefly describe the character of the cross-sections and long profiles of rivers and their valleys in the upper, middle and lower sections of their courses. (8)

8. What is meant by rejuvenation? What effects does rejuvenation produce on landforms? (11)

9. Explain how meanders are formed and destroyed. (14)

10. What factors predispose towards delta formation? Name, and quote examples, of different types of delta. (16, 17)

11. Write an essay on structure and rivers. (18, 19)

12. Explain the meaning of the following terms: corrasion, saltation, levee, pediment, knickpoint. (5, 6, 8, 9, 13)

CHAPTER XIV

THE WORK OF ICE

THE ORIGIN OF GLACIERS AND ICE-SHEETS

1. Snow, frost and ice. Reference has already been made to the action of frost in the weathering process but there are some areas in the world where frost is long-continued and its effects especially pronounced. These are regions in high latitudes and mountain areas above the snow-line. The effectiveness of frost action in the weathering process depends upon a number of variable factors including:

(a) the frequency of freezing;
(b) the availability of moisture;
(c) the character of the rock.

Areas subject to permanent frost do not suffer as much mechanical weathering as do those undergoing repeated freeze-thaw action. Rocks with, for instance, a well-developed joint system promote frost-wedging, *i.e.* water collects in the cracks or crevices, expands on freezing, and so helps to prize blocks of the rock apart. Rock-strewn areas, which are the relics of frost-wedging, are known as *felsenmeer* (literally "rock sea" or "sea of rocks").

In polar regions, and in the northern hemisphere extending generally to just south of the Arctic Circle, are extensive areas suffering from permanently frozen ground, a condition known as *permafrost*. In such areas the subsoil remains permanently frozen, even though the surface layer may thaw out seasonally.

In regions where the temperatures are below freezing point any precipitation falls as snow. If the evaporation rate and the rate of snow melting is smaller than the rate of snow-fall then the snow will accumulate to build up great masses of surface ice, for glacier ice is merely consolidated snow. Outside polar regions elevated land surfaces lying above the *snow-line* will become permanently covered with snow, *e.g.* Mt Kilimanjaro lying almost on the equator rises up sufficiently high to have a snow cap.

2. Conversion of snow into ice. When fresh snow falls it is light, fluffy and easily blown away, but after it has lain on the ground for some time it undergoes a physical change becoming compacted, heavier and granular; in this changed state it is known as *névé* or *firn*. This transformation is the result of several processes which take place, *e.g.* sublimation, compaction, melting, re-freezing, etc. As time goes on the tiny granules of ice that have been formed become increasingly locked together and the air which exists between the ice particles is gradually expelled. Finally, when the granular ice has become completely locked together into a truly solid state, it is said to form glacier ice, which is normally of a bluish-grey colour and opaque.

3. Classification of glacier ice. Glacier ice is commonly grouped into three types: valley glaciers, piedmont glaciers and ice-caps and sheets.

(*a*) *Valley glaciers.* These are streams of ice, originating in the névé fields of mountain summit areas, which flow down-slope by the easiest routes. In many respects they behave like streams of water and, as with rivers, vary greatly in their length, width, depth and speed of flow. Glaciers usually move only slowly, a few inches or feet per day although some of the Greenland glaciers, impelled by the tremendous pressure of the Greenland ice-sheet, have movement rates of as much as 50 ft (15·24 m) per day. The Alps, Himalayas, Andes and North American Cordilleras all have well-developed valley glaciers.

(*b*) *Piedmont glaciers.* Piedmont (literally "mountain foot") glaciers are formed through the coalescence of valley glaciers. As the latter leave their valleys and spread out on to the lower ground they unite to give wide spreads of very slowly moving ice. Some of these piedmont glaciers are very extensive; an oft-quoted example is the Malaspina Glacier in Alaska which covers an area of some 800 square miles (approx. 2,000 km²).

(*c*) *Ice-caps and sheets.* These are even more extensive masses of ice which spread radially outwards under their own weight. The larger spreads, *e.g.* those of Antarctica and Greenland, are commonly called continental ice-sheets while more localised spreads, *e.g.* Vatnajökull in Iceland and the Dovre Fjeld in Norway, are usually termed ice-caps. The volume of ice contained in the ice-sheets is vast; it is known, for instance, that the thickness of ice in some parts of Greenland attains 8,000 ft (approx. 2,400 m).

4. The Pleistocene Ice Age. At various times in the geological past the worsening of climatic conditions has resulted in ice ages, periods when low temperatures have led to vast accumulations of ice of continental or near-continental proportions. Geologists have distinguished a number of great ice ages since pre-Cambrian times. Each of these phases of intense cold (at least in certain areas) was separated by prolonged periods (approximately 200 million years so far as we can judge) when warmer and, sometimes, drier conditions prevailed, as in Triassic times.

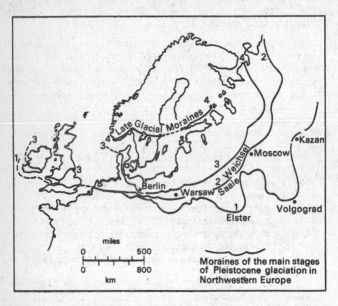

FIG. 53.—*The Pleistocene Ice Age in Europe.* The great covering of ice over most of northern and central Europe waxed and waned over one million or more years. The numbered lines mark the southernmost advances of the different phases.

The last of these periods of great glaciation occurred in Pleistocene times. The Pleistocene Ice Age probably began about 1,500,000—2 million years ago and lasted until about 8,000 B.C.; it is in fact more than likely that the small ice caps which still exist in northern Europe (in Scandinavia and Iceland) are the final remnants of the former ice-sheet which extended over the whole of north Europe. The Pleistocene ice-sheet, however, waxed and waned in its extent (Fig. 53); in other words, periods of ice advance were separated by inter-glacial phases when the ice, due to ameliorating climatic 'conditions, melted and retreated, perhaps even disappearing almost entirely except in higher latitudes and at higher altitudes. Similar ice-sheets to the one in north Europe occurred in the northern half of North America and in Siberia.

The various phases of ice advance noted above have been given different names; moreover, the European and American phases can be correlated. The names and correlations are given in Table VIII.

TABLE VIII: PHASES OF ICE ADVANCE DURING THE PLEISTOCENE GLACIATION

EUROPE			NORTH AMERICA
Alpine Region	Northern Europe	British Isles*	
WÜRM	WEICHSELIAN Eemian	DEVENSIAN Ipswichian	WISCONSIN
RISS	SAALE Holsteinian	WOLSTONIAN Hoxnian	ILLINOIAN
MINDEL	ELSTER	ANGLIAN Cromerian	KANSAN
GÜNZ			NEBRASKAN
DONAU			

*Field mapping and radiometric dating have cast considerable doubt upon the validity of some of these stages. In East Anglia, deposits previously assigned to the Anglian and Wolstonian glaciations have been shown to be the product of only one glacial period (probably the Saale) with the Hoxnian deposits post-dating them. The correlation of the Hoxian with the Holsteinian therefore seems erroneous.

5. The work of ice. The work of ice, like that of running water,
comprises erosion, transportation and deposition. Land which
has been under ice cover is always modified in one way or another
and has certain features and landforms which distinguish it
from land which has not been affected by ice action. Land which
has been altered by the erosive action of moving ice, or by the
deposition of debris from an ice cover, is called a *glaciated
landscape*.

The appearance of any glaciated landscape depends upon
two factors:

(*a*) The nature of the land that existed before the coming
of the ice.

(*b*) Whether the ice was in the form of valley glaciers or an
ice-sheet.

Mountainous country is usually shaped by valley glaciers;
this produces *highland glaciation* such as occurs in the Alps, the
Lake District or in North Wales. When the work has been done
by ice-sheets over lowland areas, such as the Canadian Shield or
the North European Plain, a very different result, a *glaciated
lowland* landscape, is obtained.

VALLEY GLACIERS AND HIGHLAND GLACIATION

6. The source of valley glaciers. When mountains rise above
the snow-line, snow collects in hollows and, as we have seen,
gradually the névé turns into granular ice and then into glacier
ice. Under the influence of gravity the ice begins to move downslope, because of its plasticity or its capacity to flow. At high
elevations, too, frost action is prevalent; the constant freezing
and thawing of water in cracks and crevices helps to shatter the
rock and sharp, angular fragments are broken off. The scree on
mountain slopes originates in this way.

The downslope movement of ice from its snowfield together
with the action of frost are responsible for erosive action. The
roughly semicircular or armchair-shaped hollows, known as
cirques (French), *corries* (Scottish), and *cwms* (Welsh), in which
glaciers originate, owe their basin-shape to the grinding action of
the moving ice and its embedded rock fragments which serve as
cutting tools although the steep walls of cirques are due to
frost action and weathering. The grinding action of the ice
over-deepens the cirque floor so that at the exit there is often a

lip or threshold. When cirques eventually become vacated by
the ice, a small lake or *tarn* is frequently left behind in the hollow
floor.

When two corries lie adjacent to each other their rear walls
are, in due course, cut back and ultimately meet to produce a
narrow, and often jagged, ridge; such knife-edge ridges are
called *arêtes* and a well-known British example is Striding Edge
on Helevellyn. If three or more cirques occur, their recession
will give rise to horn-shaped or *pyramidal peaks, e.g.* the Matter-
horn and the Zugspitze ("spitze" means spike).

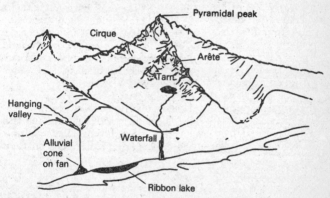

Fɪɢ. 54.—*Erosional features of glaciated landscapes.* Common features
are pyramidal peaks, cirques, arêtes, hanging valleys.

7. Glaciated valleys. A valley glacier is usually fed by the ice
from a number of cirques, in much the same way as a river is fed
by its headstreams. As the glacier moves slowly down the valley
it collects loose rock material which becomes embedded in its
icy grip and which then helps to scrape, scratch, grind and wear
away the valley sides, floors and any projecting pieces of bedrock. In effect, the glacier becomes a gigantic file. Any spurs in
the original valley are cut away or truncated while the sides are
steepened and the bottom flattened. A valley which has been
glaciated is trough-shaped (U-shaped), has long, straight
stretches, and has usually been widened and deepened (contrast
this with the V-shape of an unglaciated valley). U-shaped valleys
are common in the Highlands of Scotland and in the mountains

of central and north Wales. Glaciated valleys have irregular long profiles and sometimes long, deep, narrow lakes, *ribbon lakes*, occupy the excavated floors of such glacier troughs; Lake Windermere and Loch Ness are examples.

Hanging valleys are also features common to glaciated valleys. These are tributary valleys entering the main valley high above the valley floor, hanging above floor-level They are caused by the fact that the main valley has been eroded more rapidly than the tributary valleys, or, put in another way, the larger glacier occupying the main valley has greater erosive power than the smaller glaciers in the tributary valleys. When the ice melts away, these tributary valleys are then left hanging and streams issuing from them drop as waterfalls to the main valley floor. In the mighty ice-gouged Lauterbrunnen valley in Switzerland streams cascade and plunge several hundred feet from the hanging valleys to the main valley floor. Such falls of water are of great value for generating hydro-electric power.

8. Glacial moraine. As glaciers move down their valleys, rock-waste from scree slopes falls on their sides and this, along with material ground from the valley sides, forms *lateral moraine* (Fig. 55). If two glaciers meet the lateral moraines on their

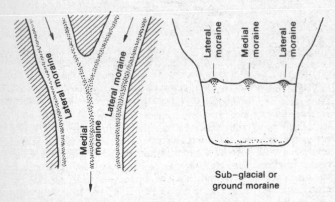

F IG. 55.—*Glacial moraines*. A bird's-eye view and a cross-section of a glacier showing the morainic debris. At the end of a glacier, where it melts, a mound of rock material, called terminal moraine, is also deposited.

inner flanks merge to form a single line of moraine known as *medial moraine*. Rock-waste eroded from the valley floor and carried along at the base of the glacier is called *ground moraine*. At the end or snout of a glacier there is to be seen a mound of debris, called *terminal moraine*, which has been deposited as the ice has melted and shed its load. Some of the finer material is carried away by the meltwater issuing from the end of the glacier.

Morainic material which is carried by and within the glacier is termed englacial moraine. That which lies on the valley floor beneath the ice, *i.e.* the ground moraine, is termed subglacial moraine, while the terminal moraine which accumulates at the snout of the glacier is alternatively known as end-moraine.

GLACIATED LOWLANDS AND THEIR FEATURES

9. Ice-sheets. We referred in **3** above to the occurrence of ice-sheets and in **4** to the Pleistocene ice-sheet which covered much of northern Europe. In Europe, as in northern North America and north-eastern Asia, the land surface was smothered in a blanket of ice which reached hundreds, and probably thousands, of feet in thickness in places. Often only the highest peaks projected above the ice, forming what are known as *nunataks*.

The ice-sheets accumulated so much ice that the weight of it was sufficient to depress the crust in those areas where the ice lay. On the other hand, these Pleistocene ice-sheets must have locked up on the land surface so much water that sea-level dropped. The present extensive ice-sheets in Antarctica and Greenland give us some sort of picture of what conditions must have been like in the various regions which were under Pleistocene ice-sheets.

The effect which this Pleistocene glaciation had on the land surface was twofold: in the areas near the centre of the ice-sheet erosive action was dominant and the ice smoothed jagged outlines, rounded hills, scooped out hollows and scraped away the soil, while areas near the edge of the ice-sheet suffered deposition.

10. Ice ploughing. As the ice-sheets grew in thickness and extended outwards, the ice-front acted like a giant scraper, ploughing all before it. The land surface was swept bare of its soil and the exposed bedrock scratched, gouged and planed down by the ice. The slow, relentless movement of the ice denuded and rounded the relief with the result that large areas of northern Finland and Scandinavia and much of northern Canada are soil-less and the bare rocks scratched and grooved (*striated*) or else polished while the sharp outlines of the topography have been gently rounded and smoothed.

Where the ice flowed over projecting masses of bedrock distinctive features were formed:

(*a*) *Roches moutonnées*: in this case the ice smoothed the "upstream" side of the rocky projection by abrasion but roughened the "downstream" side by its plucking action; rocks fashioned in this way are known as roches moutonnées. Many think that the plucking action is more important than abrasion.

(*b*) *Crag-and-tail*: in this case when the ice met an obstruction the ice was diverted over and around the rock projection and left a greatly sloping "tail" of moraine in its lee; a good example of a crag-and-tail is the crag (an old volcanic plug) on which Edinburgh Castle stands with the sloping bank of glacial debris, along which runs the "Royal Mile," forming the tail.

Boulders carried by the ice, often over considerable distances, so that they are "foreign" to their surroundings are termed *erratic blocks*, or simply *erratics*. Sometimes erratics are deposited in peculiar, balanced positions; they are then called perched blocks.

11. Ice deposition. When the ice-sheets retreated, the burden of soil and loose rock carried by the ice was left behind. Thus spreads of morainic material, termed *glacial till* or *boulder clay*, were deposited on the land surface, covering up the solid geology. (*The Institute of Geological Sciences* issues "drift maps" to show deposits of glacial material). The deposits commonly consist of a mixture of clay, sand, and boulders—hence the name "boulder clay." Till is usually unsorted material, although there may exist spreads of sand or clay which are relatively homogeneous.

Glacial deposition produces a variety of topographic forms (Fig. 56). Some areas, *e.g.* southern Finland, have a rough, irregular surface of knobby hills and deep hollows and this has been called *knob and kettle* topography. Sometimes clusters of smoothly rounded, oval hills, made of moraine occur; these small hills, about 100–500 ft (30·48–152·4 m) high, which resemble half an egg in shape, are known as *drumlins*. Their long axis reflects the direction of ice movement. Often they occur in groups or "swarms." They are common in the Vale of Eden, in the Craven Lowlands and in Northern Ireland. *Eskers* are long winding ridges, up to about 50 ft (15·24 m) high, composed of sand and gravel and they probably mark the sites of sub-glacial stream channels. Good examples occur in the plains of northern Ireland and in the Lake Plateau of central Finland. Often roads and railways have been built on the top of these natural "embankments."

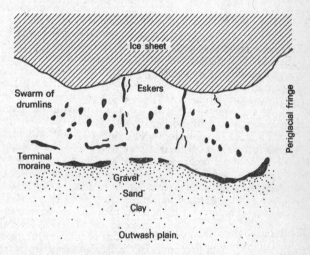

Fig. 56.—*Depositional features of a glaciated lowland*. A variety of glacial landforms are left behind by retreating icesheets including eskers, drumlins and terminal moraine. In front of the ridges of terminal moraine are to be found graded deposits *e.g.* gravels, sands and even more finely ground rock material known as rock-flour. This sorting results from melt-water washing and sorting the ground moraine.

As the ice-sheets began to melt and retreat, the melt-water washed and sorted the ground moraine. The very finely ground-up rock, known as *rock-flour*, was carried furthest away by the melt-water and relaid as a surface deposit. Rock-flour, as well as sands and gravels, is called *outwash material* because it has been washed out of the original morainic deposits. These out-washed, and to some extent assorted, materials are said to be of *fluvioglacial* origin since they are due in part to glacial action, and in part to the action of running water. A good example of an outwash plain is the lowland coastal plain of southern Iceland, termed the *sandur* plain, a fringe of clay and sand deposits.

PERIGLACIAL PHENOMENA

12. Periglacial action. The term periglacial refers to areas which are peripheral or marginal to ice-sheets and ice-caps where constant freeze-thaw action tends to be dominant. Frost-riving, solifluction and niveo-fluvial (*i.e.* snow and melt-water) processes have given rise to a variety of deposits and landforms. *Frost-riving* is the process by which water, percolat-ing through the joints, bedding planes and pores of rocks, freezes, expands and so assists in their physical disintegration through its mechanical action. *Solifluction*, which we have already referred to in XII 7(*b*), is the process by which gravity affects water-saturated, loose, unconsolidated mantle material which is transported downslope to form various features such as spreads, lobes and ridges of soil and rocks. Snow and melt-water action has been responsible for the crea-tion of overflow channels and some dry valleys.

13. Periglacial features. Features characteristic of permafrost regions are cavities formed by ice wedging, stone polygons and stone stripes, pingos and associated phenomena. Sometimes vertical cracks appear in the surface layers; these, during the short summer period when there is plenty of melt-water about, fill with water which subsequently freezes as the temperature drops in winter. Successive freezing wedges the crack further and further apart and some of these cavities may be quite deep and wide. Loose material may be washed into and fill these openings and this material preserves a cast of the original ice wedge (Fig. 57). Such casts are by no means uncommon in Britain and are to be found in the periglacial zones of the last glacial phase in Britain. Repeated freezing and thawing gives rise to frost heaving and the lifting of loose stones to the surface. In flat areas the stones show distinctive patterns and *stone circles* and *stone polygons* are formed; where the land is sloping,

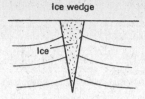 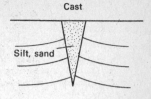

After D. Q. Bowen

FIG. 57.—*Ice wedging.* This results from water filling vertical cracks in the ground and freezing. As the water freezes to ice expansion takes place and the crack is thereby enlarged. Successive freezing over several years may produce large cavities many feet deep and wide. When the ice plug melts in summer, loose material may be washed into the cavity and this deposited material may preserve a cast, as it were, of the ice wedge.

the lifted stones form *stripes*. Stone patterns of these kinds are to be seen in the higher parts of Britain such as the Lake District and the Scottish Highlands. In many tundra areas there are frost "blisters" and associated features such as pingos; all are due to frost or ice heaving. *Pingos* are domed earth mounds. They are caused by water in the soil freezing, expanding and lifting the surface material up into a dome-shaped mound. When the core of ice melts the pingo collapses to leave behind a depression, sometimes partially filled with water, enclosed by a hummocky earthen rampart.

PROGRESS TEST 14

1. In which areas of the world is the action of frost, snow and ice of particular significance? And which are areas subject to permafrost? (1)
2. Suggest a threefold classification of glacier ice and write a short paragraph on each type. (3)
3. Give an account of the duration, occurrence and phases of the Pleistocene glaciation. (4)
4. Explain the following terms: névé, felsenmeer, permafrost, cirque, erratic. (1, 2, 6, 10)
5. What glacial features can be seen in such glaciated regions as the English Lake District, Snowdonia, the Alps or the Canadian Rockies? (6)
6. What is moraine and what different types of moraine may be distinguished? (8)
7. What are the characteristic features of glaciated valleys? (7)
8. Describe the landscape effects of "ice ploughing." (10)

9. What different kinds of deposits and depositional features are to be found in regions of ice deposition? **(11)**

10. Draw annotated diagrams to illustrate the following landforms associated with, or resulting from, glacial action: nunataks, cirques, drumlins, roches moutonnées. **(6, 9, 10, 11)**

11. Explain the meaning of the term "periglacial." What are the various processes associated with periglaciation? **(12)**

12. Describe the various features which may be seen in periglacial landscapes. **(13)**

CHAPTER XV

THE WORK OF THE WIND

WIND AND WEATHERING

1. Wind action. We are all familiar with the destructive capacity of galeforce winds: hurricanes and tornadoes, even high winds in Britain, are capable of doing an enormous amount of damage. Wind by itself, however, though capable of uprooting trees and of blowing roofs off buildings, does no erosive work unless it possesses cutting tools in the form of fine particles of rock. In this sense, the wind is rather like running water which, without its load of silt and pebbles, undertakes very little erosive work; although water can dissolve matter, something the wind is incapable of doing.

The wind is an important geological agent, although much of the work which formerly was attributed to wind action is now recognised as being the result of water action; in other words, there have been changes in climatic conditions and much of the erosion now seen in desert areas took place in earlier pluvial periods.

The work of the wind is most effective in arid regions, *i.e.* in dry climates where the rainfall is small (approximately under 10 in (25 cms) or where nearly all the rain that falls is evaporated. Generally speaking, *wind action is the most important denudational force in arid lands*. The action of water is of subordinate importance, although even small amounts of moisture may play a very significant role in the fashioning of arid landscapes. Occasionally torrential rains, due to thunderstorms, give rise to sudden floods which carry out the work of erosion and deposition speedily.

NOTE: in humid regions, especially where the ground is covered with vegetation and where weathered material is held together by moisture, there is little evidence of the work of the wind apart from coastal sand-dunes.

2. Weathering. The actual work of the wind is greatly helped by the weathering process. In arid lands clear skies allow the sun to beat down strongly and there are great daily ranges of temperature. These two conditions assist weathering. The sun heats up the rock surfaces and causes the mineral grains of which the rock is composed to expand. After sundown the land cools down rapidly by radiation and the heat escapes because the night skies are clear. The constant expansion and contraction of the mineral particles gradually weakens the rock.

Furthermore, heat penetrates rock only slowly, so that while the outer surface is hot the inside remains cool. Thus there is a tendency for the outer surface to prise itself off from the interior. These mechanical actions of mineral and rock expansion and contraction lead to the shattering and flaking of the rock. Rocks often show a scaly appearance where layers have peeled off. This onion-like peeling is known as *exfoliation*. Alternatively, the rock becomes shattered and angular fragments split off, often piling up to form scree.

Although arid regions have little rainfall, there is nearly always *some* moisture present, if only in the form of dew. This moisture assists the breakdown of the rock by helping chemical processes to take place. Many desert rocks show evidence of rotting which can only have occurred through the presence of moisture.

Thus weathering, due to temperature extremes, chemical activity, and the absence of a protective vegetation cover, occurs even if only slowly and provides plenty of loose material for the wind to move and work with.

DENUDATION BY WIND

3. Wind erosion. Two kinds of work are involved in wind erosion:

(a) *Deflation*, the lifting into the air or the rolling along the ground of loose particles of rock, such as dust, grains of sand and pebbles.

(b) *Abrasion*, the sand-blast action which the wind has on rock surfaces when it hurls sand against them.

Deflation is the work of air currents alone; abrasion needs windborne cutting tools.

4. Deflation. Deflation may be defined as the aspect of wind denudation which is concerned with the removal, by blowing away, of loose surface material. Wind action sorts out loose, unconsolidated surface material:

(a) It picks up the minute dust particles and blows them away.

(b) It rolls or "jumps" along—a process known as *saltation* —sand and small rock fragments.

(c) It leaves behind pebbles and large boulders it is unable to move as residual material.

Because of this selective action, various kinds of desert surface are produced, *e.g.* sheets of pebbles or angular rock fragments are termed *reg*, piled-up sand forming dunes is known as *erg*. while bare rock surfaces form *hamada*.

Sheets of rock fragments are often closely fitted together and wedged piece with piece; such mosaic-like spreads from which all the fine grains have been removed are called *desert pavements*.

Deflation over a prolonged period may result in the excavation of wide, though usually fairly shallow, basins or depressions, sometimes termed *blowouts*. They may be of various sizes but one, Big Hollow, in Wyoming in the United States, is of exceptional size: it is 9 miles (14·4 km) long, 3 miles (4·8 km) wide, and 300 ft (90 m) deep. Blowouts are typical of arid plains regions. Where such excavations reach downwards and cut the water-table, they form oases.

5. Abrasion.
Strong winds armed with sand have a sand-blast action. They pick out all the points and lines of weakness in the rock and often carve it into fantastic and grotesque shapes. Cliff faces, for instance, may be pitted to such an extent that they look like lace curtains; such fretted faces are termed *rock lattices*.

Wind driven sand seldom rises to much more than three or four feet (0·9–1·2 m) above ground level and most of the sand is concentrated in the eighteen inches (450 mm) nearest the ground; hence the sand-blast effect is greatest at or near ground level. As a result, upstanding rocks are undercut at their bases to produce *pedestals* or *mushroom* rocks (Fig. 58).

Where rocks of differing hardness are horizontally-bedded, features known as *zeugen* may be formed. If the hard capping rock becomes cracked or broken, the wind undertakes abrasive action along such points of weakness until finally it cuts through the hard layer into the softer strata below. Continuing erosion produces tabular masses of resistant capping rocks perched upon the softer rocks beneath. These upstanding and usually flat-topped, straight-sided ridges or zeugen often run in parallel sequence. They are a feature of the "badlands" areas of the United States.

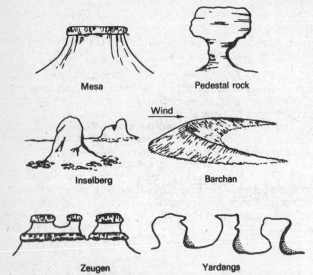

Mesa

Pedestal rock

Wind

Inselberg

Barchan

Zeugen

Yardangs

FIG. 58.—*Features of an arid landscape.* A variety of erosional and depositional features associated with dry lands where wind action is a pronounced agent of denudation and deposition.

In central Asia a similar wind abraded feature is the *yardang*; these appear to result when alternating hard and soft beds outcrop vertically. The upstanding hard rock ridges, often fantastically shaped, are separated by wind-cut troughs. These features again show a roughly parallel alignment and run in the direction in which the prevailing wind blows.

Ventifacts, or wind-abraded pebbles, are common features. Pieces of rock, which are too heavy to be moved by the wind or which have become wedged in the ground, become abraded on the windward side and the surface planed down and polished. If the wind changes direction or if the rock fragment is moved, new faces will be presented for abrasive action. A common kind of ventifact is the *dreikanter* or three-faced pebble.

WIND TRANSPORTATION

6. Means of transport. Rock particles are moved by the wind in three different ways:

(a) *By suspension:* dust, very fine particles with diameters of less than 0·06 mm, is picked up by the wind and held aloft by air eddies and may be carried over long distances.

(b) *By saltation:* sand grains are usually too large to be held aloft for long and they tend rather to be moved along in a series of jumps; when a grain of sand falls, it may strike another and ricochet off it; thus grains of sand are impelled forward by constant impact and rebound.

(c) *By surface creep:* larger grains of sand never become airborne, they merely roll forward; strong winds blowing over flat surfaces may roll pebbles up to 2–3 in (50–75 mm) in diameter along the ground.

It is generally believed that about three-quarters of the drifting sand travels by saltation, the remaining quarter moving by surface creep.

7. Quantities moved. It is extremely difficult to estimate the amount of material transported by the wind in any year, but it is beyond any doubt very large. A number of estimates have been made and these, even if they err on the generous side, indicate that the amount of material carried by the wind annually must be enormous, for example:

(a) one Saharan dust storm which covered nearly half a million square miles (1,295,000 km^2) resulted in an estimated 2 million tons of dust being deposited over an area of 168,500 square miles (436,400 km^2) (Worcester);

(b) a dust storm in Nebraska, in the United States, during the "dust bowl" period of the 1930s, resulted in a deposit of 35 tons per square mile (14 tons/sq km);

(c) dust storms in one year in the western part of the United States led to an estimated 850 million tons of dust being carried over a distance of 1,500 miles (2,413 km) (Udden).

WIND DEPOSITION

8. Dunes. Sand is merely rock waste which has been pulverised into small grains. In many arid areas (and also in some coastal districts of humid areas) sand is accumulated in quantity by the wind and is also shaped by it. Such mounds of sand are called dunes. There are two main types of dune:

(a) *Barchans*, crescent-shaped mounds, usually occurring in groups.
(b) *Seifs*, long parallel ridges of sand.

Dunes vary widely in size: they range from only a few feet in height and a few yards in length to as much as 700 ft (213 m) in height with base widths of 2,000–3,000 ft (609–914 m).

9. Barchans. The crescent barchan possesses two horns or wings which point down-wind. To windward the dune presents a gentle slope up which the sand grains are blown; on reaching the crest, they roll down the lee slope which is comparatively steep. Because of this windward sweeping and leeward accumulation the dunes are constantly migrating down-wind.

The dunes develop in level areas. Occasionally barchans occur singly but more commonly they occur in vast colonies where the dune ridges are partially coalesced.

10. Seifs. Seif dunes are long and relatively narrow ridges of sand, sometimes running for miles; they get their name from their fancied resemblance to the long, flat blade of an Arab sword. Seifs owe their elongated shape to dominant winds which blow, alternately, from one quarter of the compass to another adjacent quarter.

The formation of seif dunes is not completely understood but it is thought they originate from barchans (Fig. 59). If a cross-wind blows a barchan will begin to swing round and one of its horns will become elongated. If the wind from the new direction persists, then the extended horn will continue to grow and in due course the barchan will change its shape and become altered into a seif dune. Some authorities believe seifs are formed by a barchan becoming breached in its central part.

11. Loess and kindred deposits. Wind, as we have already noted, is capable of carrying dust over great distances. Dust from the Sahara, for example, is often blown northwards across the Mediterranean Sea and colours the rainfall producing what are known, in southern Italy, as "the blood rains." In eastern

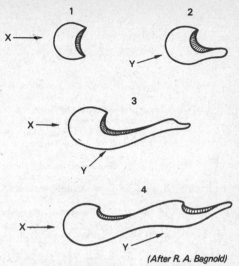

(After R. A. Bagnold)

Fig. 59.—*The formation of a seif dune.* The formation of seif dunes is not completely understood but it is believed they originate from barchans. Some believe persistent cross-winds cause the barchan to swing round thereby leading to the elongation of one of the "horns"; others think seifs may be due to barchans being breached in their central parts.

Asia the winter monsoon winds have, over a period of many thousands of years, blown vast quantities of dust from the Gobi Desert over northern China where it has settled and blanketed the landscape, in some places to depths of several hundreds of feet. This fine, calcareous, buff-coloured, non-stratified deposit is called *loess*. On the plateaus of northern China the valleys and lower hill slopes are smothered by loess, but the hilltops rise above this mantle of loess like islands in the sea. The deposit is soft but, at the same time, resistant; hence rivers like the Hwang-ho have carved deep narrow canyons in it.

A similar deposit of wind-blown origin (but probably also owing something to ice and river action) occurs on the southern margins of the North European Plain and on the northern lower flanks of the Hercynian ranges; in Germany this deposit goes under the name of loess, which is a German term, but in the

Paris Basin it is called *limon*. The *adobe* deposits of the Missouri–Mississippi Basins are probably very similar. Similar loess like spreads occur in Argentina.

PROGRESS TEST 15

1. In what ways is wind action (*i*) like and (*ii*) unlike the action of running water ? (1)

2. How does the weathering process assist wind action ? (2)

3. What different kinds of work are involved in wind erosion ? (3)

4. Describe the various ways in which wind undertakes deflation. (4)

5. Describe the abrasive action of wind and some of the landforms resulting from it. (5)

6. Explain the following terms: exfoliation, hamada, ventifact, rock pedestal. (2, 4, 5)

7. Wind removes rock particles in three different ways: what are these ways ? (6)

8. Describe the features and explain the forms of barchans and seif dunes. (8, 9, 10)

9. Describe the origin, nature and occurrence of loess and other associated deposits. (11)

10. What are the following and where are they to be found: Big Hollow, yardangs, "blood rains," adobe ? (4, 5, 11)

CHAPTER XVI

THE WORK OF THE SEA

THE EROSION OF THE COAST

1. The coastline. Between John o' Groats (the most northerly point of the mainland of Britain) and the Lizard (the most southerly point), the coast has great variations in its shape and form: it includes sheer cliff faces and flat, featureless shores, rugged, highly broken coasts and straight, gently shelving beaches, majestic white chalk cliffs and dismal, muddy estuarine flats. But whatever the configuration and character may be, the coastline is a zone of great interest and attraction as the millions of holiday-makers who wind their way to the seaside each year show.

The coastline, like the land surface, is constantly undergoing change. Such changes are usually slow and imperceptible, although occasionally, as following a severe storm, we can see the havoc wrought by the sea in its angry moods: promenades may be damaged or piers broken, cliffs may collapse or shingle be thrown up in great piles, while the land may be invaded by the sea, which has burst through the sea defences, and large areas flooded, as happened along the coast of eastern England and in the Netherlands during the great storm of 1953.

Along any stretch of coast we can, broadly speaking, observe two very different activities of the sea: one is destructive, where erosion is taking place, the other is constructive, where deposition is occurring. In general, erosion at any point is balanced by deposition at another, so that we might describe the work of the sea as a striving to reach some sort of coastal equilibrium.

2. Coastal erosion. The chief erosive forces fashioning the coasts are the waves of the sea, assisted by tides, currents and such ancillary aids as the wind, sea spray and rain. Waves, especially storm waves, have enormous power: they can lift and displace huge rocks or blocks of concrete and as they smash against the cliff face they are capable of dislodging and removing pieces of rock. As the waves hurl particles of sand and beach boulders against the base of the cliff, the scratching, scraping and battering weakens and wears away the rock. The impact of boulders helps to crack and break the cliff face: at the same time, the boulders themselves may be shattered into pieces. The continual breaking up and fragmentation of beach material (boulders, shingle) is called *attrition*.

The disintegration of cliffs is also undertaken in another way: as waves crash against the cliff face they compress the air in the cracks and crevices of the rock; as the waves retreat the pressure is suddenly released. The sudden and constant alternation of the contraction and expansion of the air, weakens the rock structure and helps to loosen particles of rock, which eventually are broken off.

The constant swish-swash and pounding of the waves, with their suspended cutting tools of sand and pebbles, at the base of the cliff acts like a file and, slowly but surely, cuts into the cliff along a narrow zone between the high and low tide-marks. In time, a fairly smooth wave-cut platform, called an *abrasion platform*, is formed. Good examples of these wave-cut platforms can be seen along the north-east coast of Yorkshire in the Flamborough Head and Robin Hood's Bay areas.

3. Factors in coastal erosion. The amount of erosion which takes place depends upon several factors chief of which are:

    (*a*) the degree of exposure to wave attack;
    (*b*) the degree of resistance of the rock;
    (*c*) the structure of the rocks;
    (*d*) the effect of tides and coastal currents;
    (*e*) the amount of protective interference by man.

If the land faces open water the wind can build up larger waves and the rougher the sea the greater is its destructive effect. During storms waves travel farther, reach higher, and may excavate great hollows along the coast. Storm damage can be great: we have all seen photographs of yawning gaps in promenades torn out by angry seas, and some years ago at Wick in the north of Scotland part of a massive breakwater, weighing some 2,500 tons, was moved out of position during a storm.

Soft rocks, such as clays and some sandstones, are easily worn away and clay cliffs, if moistened, may be subject to landslips.

Such landslips are especially liable to occur if the beds of rock dip towards the sea, for undermining by the waves at their base will cause them to fall forward. Hard rocks are usually more resistant to attack but some hard rocks such as granites and some limestones, which often have joints and cracks in them are liable to penetration and erosion by the sea along such lines of weakness. Erosion at these weak points leads to the formation of deep, narrow clefts, called *geos*, which are familar features along the North Yorkshire coast and in Pembrokeshire, and to the formation of caves (Fig. 60). Sometimes *blowholes*

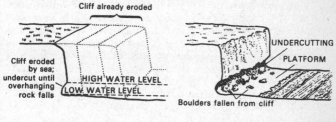

FIG. 60.—*Features of coastal erosion.* The top two figures show how a wave-cut platform is produced. The lower diagram shows various coastal features (geos, caves, blow-holes, arches, stacks) which are produced by marine erosion.

or *chimneys* are formed in the roofs of caves: the constant splashing of sea-water against the cave roof, as waves hit the back of the cave, results in the erosion of the joints above; eventually a hole is driven through the cave roof, and spray occasionally may be seen to rise through the aperture. Where hard rocks alternate with soft rocks, as in the Tor Bay area, differential erosion takes place: the resistant rocks jut out as headlands and peninsulas, while the softer rocks have been eroded to form bays.

4. Cliff morphology. The form which cliff faces take is determined primarily by the type or types of rock of which they are composed and the structure of the rocks. Where soft rocks occur, as in the boulder clay cliffs of Holderness, steep faces are characteristic as a result of constant under-cutting. Moreover, the eroded boulder clay provides an abundance of debris which assists marine erosion. Where softer rocks underlie harder ones, undercutting action weakens and finally removes basal support which results in slope failure. In the Flamborough area, where boulder clay lies on top of the chalk, subaerial activity leads to constant slumping; hence the cliff profile is more or less vertical in the chalk, but becomes sloping in its upper part.

Massive rocks are usually more resistant than bedded rocks. The dip of bedded rocks is of some importance. Cliffs formed of horizontal or vertical beds tend to have steep profiles; where the strata dip landwards, stepped cliffs are often formed; where the strata dip seawards, overhanging cliffs often result, although the cliff profile is largely controlled by the angle of dip. Many high, vertical cliffs are the outcome of faulting, e.g. plunging cliffs which descend into deep water.

5. Arches, stacks and stumps. Many headlands show marine features which represent evolutionary stages in the development of such landforms. Due to refraction wave attack tends to concentrate upon headlands, hence caves are sometimes formed on either side of a promontory. If such caves happen to be situated more or less back to back, eventually the rear wall of one cave will break through into the adjacent cave. When this happens an *arch* is created, such as the Needle Eye near Wick. In time this arch will collapse leaving pillars of rock, or *sea-stacks*, standing in coastal waters; the Old Man of Hoy, in the Orkneys, which is 450 ft (137 m) high, is a particularly fine example. In due course, these rock pinnacles crumble and collapse and end up as mere *stumps* standing only slightly above sea-level. All these features—caves, arches, stacks, stumps—are merely stages in the general process of cliff retreat.

6. The rate of coastal erosion. The rate at which a coast wears away or retreats depends very much on the combination of factors noted in **3** above. Tough, hard rocks which are highly resistant to erosion, e.g. granite, wear away only slowly. Conversely, soft rocks, such as clays and shales, may be eroded relatively quickly, especially if they are exposed to strong wave attack and strong currents move alongshore. Many parts of the west coast of Scotland probably lose less than 1 in (25 mm) per century. Quite the opposite is the Yorkshire coast between Flamborough Head and Spurn Point; here the coast is largely made up of soft boulder clay (lying on chalk) and erosion is taking place here at a higher rate than anywhere else in the British Isles. The coast is receding by between 5 and 7 ft (1·5 and 2 m) per year and since Roman times a strip of land, approximately 2 to 3 miles (3 to 5 km) in width, has been eroded away. Numerous villages and towns which once existed along the coast of Holderness now lie beneath the waves.

Where coastal erosion is serious, man has built a variety of sea-works and defences, such as groynes, breakwaters and concrete walls, to help check erosion. Marine civil engineering, however, is often a very expensive business.

COASTAL DEPOSITION

7. Depositional landforms. A variety of distinctive landforms has been built up around coasts by marine processes: some, as noted in the above, are the result of erosive, destructional action but others are the outcome of deposition. Depositional features are not as dramatic as the landforms created by marine erosion; the former are, nevertheless, widespread and varied in their character. Some coastal depositional features, such as deltas, result largely from river action but here we are concerned only with those formations resulting from marine agencies, e.g. beaches, bars, spits, tombolos and cuspate forelands.

In general terms, coastal depositional formations are composed of either sand or shingle, or both, which are thrown up by wave action above highwater mark: in other words, they are essentially features constructed by wave action, though sea currents and drifts, wind and plants may also play a supporting role. "Coasts of deposition are made up of three elements: the wave-built ridge or bar of sand and shingle, sand dunes, and salt marshes. But clearly it is the wave-built structure which is of primary importance and upon which the others, if they are present, depend" (G. de Boer, "Land created by sea", *Geographical Magazine*, Feb. 1973, p. 379).

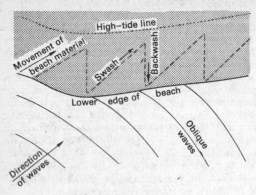

(a) LONGSHORE DRIFT

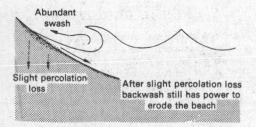

(b) ACTION OF STEEP DESTRUCTIVE WAVES

After G. de Boer

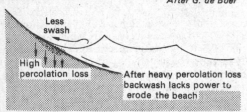

(c) ACTION OF FLAT CONSTRUCTIVE WAVES

After G. de Boer

FIG. 61.—*Wave action.*

8. Wave action. The action of waves is largely responsible for the creation of marine depositional features. Waves may be either destructive, as we have seen in **2** above, or constructive (Fig. 61). Whether they are destructive or constructive in their effects depends upon their steepness. Where steep waves roll up the beach the *swash* or uprush of water runs well up the beach but the *backwash* is vigorous and powerful enough to carry beach material down-beach. On the other hand, where the incoming waves are flatter, the swash is less abundant but the backwash is also much less powerful and lacks erosive power; hence, more material is carried up-beach by the waves than is carried back to the sea. *Flat waves then are constructive waves.* It should be understood, however, that any stretch of coast will receive both steep and flat waves depending on the wind and the weather, but whether erosion or deposition predominates will be related to whether steep destructive waves or flat constructive waves are dominant.

9. Marine transport. Just as rivers perform a threefold function—erosion, transport and deposition—so also does the sea. Marine transport is effected by both waves and currents.

The material is eroded from the coast by waves in two main ways:

(*a*) Material is carried seawards, as we have just noted, by backwash and undertow. Just as waves cause a slight piling up of water against the coast, so a compensating current carries water seawards along the sea-bed. Sea bathers are often conscious of this pull exerted by the *undertow*. The undertow has the effect of dragging material out to sea.

(*b*) Waves usually travel obliquely on to the shore, hence, bit by bit, they transport sand and shingle along the shore in the direction of their movement. Note that though the swash moves obliquely up the shore, the backwash from the spent wave moves directly down the beach slope to the sea. This transport of beach material along the shore is called *longshore drift*.

Much transport is also carried out by currents and tidal streams. Currents have neither the power nor the energy of waves and so they can carry only the finer material eroded from the land. But the finer particles, which have been carried by the waves and deposited as silt in the shallow waters near the coast, are disturbed and picked up by coastal currents and transported farther along the coast. And it is clear from the vast accumulations of sand and shingle which make up shoals, bars and spits that enormous quantities of eroded material are transported by wave and current action.

10. Coastal deposition. Frequently, at various points along a coastline, the transporting power of the waves or the longshore movement of drift is interrupted or obstructed in some way, and this results in deposition taking place at these particular points. This is likely to occur where, for example, rivers enter the sea, where tidal currents neutralise each other, or where embayments in the coast provide areas of sheltered water. Any kind of check to water movements and, therefore, to the carriage of either beach material or the suspended load of silt and sand is likely to cause deposition and the building up of coastal landforms.

11. Bars. At Loe, near Helston, in Cornwall and Slapton Sands in Devon, *bars* have been built across the mouths of rivers. These bars, running smoothly across the estuary mouths, have cut the river off from the sea and led to the formation of Loe Pool in the first case and to a lagoon in the second. Such bars are usually made up of river sediment on their landward side and of shingle to seaward. Lagoons thus enclosed are temporary features for, in time, they become filled in by waves washing beach material into the lagoon and by the gradual accumulation of sediment brought down by the rivers which is unable to escape out to sea. The river-borne silt settles behind the bar and builds up to form *mudflats* which, as they become colonised by distinctive plant communities, develop into *salt marshes*.

The shingle ridge at Slapton Sands is of special interest since it is made up almost entirely of flints from chalk which is not to be found anywhere in the vicinity. How can this be explained? It has been suggested that the material has been washed up from the sea-bed where glacial drift deposits had been dumped during the Ice Age.

12. Spits. Sometimes deposition takes the form of *spits*, long narrow accumulations of sand and shingle which grow outwards

from the land to which they are attached and project into the
sea. They commonly grow across the mouths of estuaries or
bays. Spurn Point, at the mouth of the Humber, provides a very
fine example of a spit. Here, large quantities of material, eroded
from the soft boulder clay coast of Holderness, are carried
southwards by a southward-flowing current. When this current
meets the current of the Humber, waterborne material is
jettisoned and so a great spit has grown up. Many complex
factors are responsible for the formation of Spurn Point, but in
essence it has grown up as a result of the obstruction of longshore
drift.

Many wonderful sandspits, in various stages of growth, are
to be found along the southern shore of the Baltic, where often
they enclose extensive lagoons; these spits and lagoons here are
known as *nehrungs* and *haffs* respectively. Figure 62 shows another example of a spit.

13. The rate of marine deposition. As with erosion, marine
deposition varies greatly. Sometimes it is surprisingly rapid.
Along the Norfolk coast, for example, there are many small
towns which two or three hundred years ago were ports on the
coast, but which now lie more than a mile (over 1·5 km) inland.
At Southport, on the Lancashire coast, what is now the main
street lay under water at high tide in 1736, and the sea is receding from this "seaside" resort so rapidly that a railway was
built on its very long pier to enable holiday-makers to get a
glimpse of the sea! Further north in Morecambe Bay there are
extensive areas of salt marsh, the result of deposition, and at
Silverdale one can see the former sea cliffs fronted by wide
stretches of "new land."

OCEANIC DEPOSITS

14. Classification. Just as soil covers the land surface, marine
deposits of mud, sand, clay and ooze cover the ocean bed. These
deposits are classified into two broad groups based upon their
origin, composition and the depth at which they occur:

(a) *Terrigenous* if derived from the land.
(b) *Pelagic* if derived from marine organisms.

This distinction is based upon the fact that the nature and

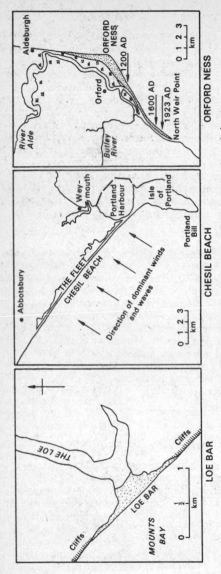

LOE BAR   CHESIL BEACH   ORFORD NESS

Fig. 62.—*Loe Bar* is a simple bar built across the mouth of a river; it is made up of river sediment on its landward side and of shingle to seaward. Chesil Beach is a great shingle ridge built up by wave action and long-shore drift. Orford Ness is an alluvial and shingle spit; a southward flowing current helped to carry river alluvium and beach material southwards to create a long tapering spit.

composition of the material covering the sea-bed near to land differs completely from that found on the floor in mid-ocean.

15. The terrigenous deposits. These may be sub-divided into two types:

(a) The sand, shingle, gravel and shelly remains found on the continental shelves.

(b) The fine-grained muds of various colours which occur on the continental slopes beyond the hundred-fathom (183 m) line.

Both these groups of deposits are derived from the land. The material shows a gradation in texture with distance from the shore: the fine muds and clays lie farthest from the land and in deep water. The muds are usually classified as *blue, green* and *red muds*; their colouring is due to chemical constituents, especially ferruginous material, in them.

Away from the continental margins, land-derived material becomes more and more rare and, usually, at a distance of some 200–300 miles (320–480 km) offshore the terrigenous deposits become replaced by pelagic deposits. (*see* Fig. 63).

16. The pelagic deposits. These comprise two different types: ooze and clay. The ooze consists of the shelly and skeletal remains of microscopic marine organisms. Countless millions of these tiny creatures live in the ocean; as they die they sink to the ocean bed and their remains slowly accumulate. The residual deposit which is built up is usually greyish in colour; when dry, it is a little like flour in texture. The deposits of ooze consist mainly of either carbonate of lime or silica and it is largely on the basis of composition that the different types of ooze are recognised. Four main types occur:

(a) *Globigerina ooze* ⎫
(b) *Pteropod ooze* ⎬ composed of calcium carbonate.

(c) *Diatom ooze* ⎫
(d) *Radiolarian ooze* ⎬ composed chiefly of silica.

In the deeper parts of the ocean basins, and occurring over wide areas of the Pacific in particular, is a stiff brownish-red clay; this deep-sea deposit is known as *red clay*. In these deeper parts of the oceans, the oozes are absent since the calcareous

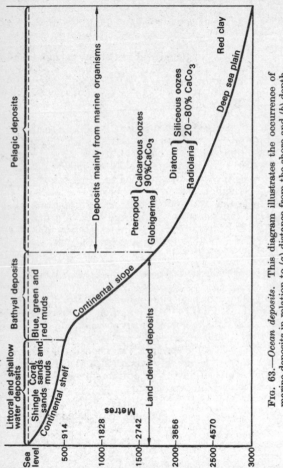

Fig. 63.—*Ocean deposits.* This diagram illustrates the occurrence of marine deposits in relation to (*a*) distance from the shore and (*b*) depth of water.

and siliceous remains of the marine organisms are completely dissolved away before they reach the ocean floor. Red clay is believed to consist of dust particles, especially volcanic dust, and is almost free from the remains of marine organisms, the only substances found being especially hard parts such as sharks' teeth and the ear-bones of whales. Red clay accumulates very slowly, perhaps only a millimetre or two per century.

PROGRESS TEST 16

1. Coastlines display great variations but all are subject to changes. Describe these variations and indicate why changes constantly take place. (1)

2. What factors are at work in the process of coastal erosion? (2)

3. The amount of coastal erosion which takes place at any point depends upon a variety of factors. What are these factors? (3)

4. Explain the evolutionary and sequential stages in the development of sea caves, arches, stacks and stumps. (5)

5. Show how the rate of coastal erosion varies between different stretches of coastline. (6)

6. What are the three elements of coasts of deposition? Name the chief kinds of depositional landforms. (7)

7. Explain how waves and currents undertake the work of marine transportation. (8, 9)

8. Explain how bars and spits are formed. Give examples of each. (11, 12)

9. What are the essential differences between terrigenous deposits and pelagic deposits? (15, 16)

10. Name the different types of pelagic deposits and note the main characteristics of each type. (16)

CHAPTER XVII

GEOLOGICAL HISTORY OF THE BRITISH ISLES

1. Introduction. Using the principles of stratigraphy geologists have been able to piece together the complicated story of the history of the earth. In this chapter we shall give a brief review of that history as far as the British Isles is concerned. It should be remembered that the islands as we know them today have only taken shape in comparatively recent times, and that the history we are tracing is really the history of a small portion of the earth's lithosphere which has drifted across the face of the earth taking part in plate closures and separations as it did so. Only the main developments can be mentioned here, but the student who is interested in the details of this evolution can refer to more advanced texts.

PRE-CAMBRIAN AND PALAEOZOIC ERAS.

2. Pre-Cambrian times. The first 3,000 million years or so of the earth's history are very difficult to trace with any degree of exactness. Most of the rocks belonging to this period have undergone so many phases of deformation that they reveal very little of their conditions of formation. Yet they must represent many momentous phases of earth evolution, *e.g.* the development of the first oceans, the formation of life, the change in the atmosphere from a reducing to an oxidising one, etc.

The most extensive outcrop of the Pre-Cambrian in the British Isles occurs in the North West Highlands of Scotland and extends across into Northern Ireland. Minor outcrops are also found in Anglesey and North Wales, South Wales, the Church Stretton area, the Malvern area and Charnwood Forest.

The Scottish Pre-Cambrian strata have been divided into four major series. The *Lewisian* and *Torridonian Series* are located on the western fringe of the Highlands to the west of the Moine thrust zone. The *Moine* and *Dalradian Series* occupy the rest of the Highlands to the east and south. By far the oldest of these series is the Lewisian which consists of highly altered gneisses cut by a group of altered basic intrusions. Radiometric dating suggests that the periods of metamorphism which affected those rocks occurred in some instances 1,600 million years ago (Laxfordian episode) and in others as far back as 2,700 million years ago (Scourian episode).

In contrast to the Lewisian, the Torridonian consists mainly of relatively little altered sandstones and siltstones. They sit unconformably upon the Lewisian and do in fact bury a former Lewisian landscape. In places they cover former hills which were several hundred feet high. Sedimentary structures such as current bedding, ripples, desiccation cracks, etc., suggest that the Torridonian beds in the north are fluviatile and intertidal, whilst those in the south are more typical of deeper water marine sedimentation. Thus a picture is obtained of an ancient Lewisian landmass (now in fact part of North America) lying to the north and west with an oceanic area lying to the south.

It was in this ocean that the Moinian and Dalradian accumulated. The Moines are most likely to be roughly equivalent to the Torridonian in age (relatively late Pre-Cambrian) although they differ in having been deposited further from the ocean margin. Moreover they have been metamorphosed and now appear as a series of schists. Unlike the Lewisian, however, their metamorphism took place during the Caledonian orogeny long after Pre-Cambrian times. The same is true of the Dalradian Series, originally a group of sandstones, shales, and limestones, but now appearing as a rather monotonous sequence of schists. Trilobites found near the top of this sequence indicate a Lower Cambrian age and thus the Dalradian possibly spans the Pre-Cambrian–Cambrian boundary. A boulder bed in the Dalradian (Schiehallion boulder bed) has been interpreted by some geologists as a tillite. This along with similar evidence from Scandinavia and Greenland suggests that there was a late Pre-Cambrian glaciation in these areas. Palaeomagnetic evidence would tend to fit in with this, for it suggests that at this point in time the area under consideration lay in the high latitudes of the *southern* hemisphere.

3. Lower Palaeozoic times. The Lower Palaeozoic comprised the Cambrian, Ordovician and Silurian periods. The rocks of these systems are found predominantly in the west and north of the British Isles. They occupy large areas of Wales, the Lake

District and the Southern Uplands, with continuations of these belts in Eastern Ireland. In addition there are smaller outcrops in the North West Highlands of Scotland, Shropshire and the Midlands, and North West Yorkshire.

The strata reveal the history of the ocean that has already been discussed in connection with the Moine and Dalradian sediments. The great thickness of sediments which were

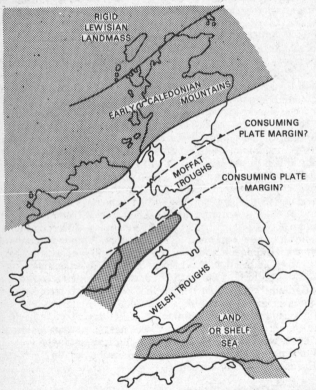

FIG. 64.—*Simplified lower Palaeozoic palaeogeography*. Note that in this figure and Figs. 65—73 the outlines of the British Isles and the various continents are shown for reference purposes only. They did not exist in this form during the periods represented by these maps.

collecting in this ocean lead us to believe that it was part of a great trough stretching from Scandinavia through Britain to the Northern Appalachians (it must be realised that Britain and America were not in the same relative positions or of the same geography as they are today). This trough is termed the *Lower Palaeozoic or Caledonian Geosyncline or Ocean.*

The Cambrian beds give the first indication of a subsiding sedimentary basin or trough in the Welsh area, for in the Harlech Dome a thickness of around 3,500 metres of grits and shales is recorded. Moreover, the thinning out of these sediments both to the south-east and north-west suggests that this trough was not a straight-forward continuation of the Dalradian one further to the north. Indeed, the succeeding Ordovician and Silurian beds confirm this belief and indicate that the Caledonian geosyncline consisted of a series of developing troughs with geanticlinal rises between them (Fig. 64). It appears probable that these shifted position with time.

The Ordovician and Silurian beds also show that tectonic and volcanic disturbances were occurring. In Wales, the Lake District, and the Southern Uplands the rocks are mainly greywackés, black shales, or volcanics; in the Welsh borderlands and Midlands, sandstones, limestones and mudstones with a shallow-water shelly fauna predominate. The latter indicate the south-eastern limit of the troughs. Widespread unconformities in the sediments suggest tectonic unrest and studies of the volcanic rocks supply evidence that a number of volcanic islands were building up out of the sea. This is taken to imply that a southern margin subducting plate boundary existed just to the north of Wales and the Lake District; a further one may also have occurred on a line which now lies along the northern side of the Southern Uplands. This would have been on the northern margin of the Caledonian ocean.

It is likely that these events were heralding the closing of the Caledonian ocean, as a plate, which included parts of Europe, moved northwards crushing the sediments against a plate containing parts of North America, Greenland and North West Scotland. It is interesting to note in this connection that the Pre-Cambrian (Torridonian), Cambrian and Ordovician (Durness Beds) sediments and fauna of the north-western fringe of Scotland are very similar to those in the Pacific Province of North America and quite unlike those of the Anglo-Welsh zone to the south. This suggests that these two faunal types evolved on opposite margins of what must have been at one time a fairly wide Caledonian ocean.

4. Devonian times. The final phases of the closing of the geosyncline produced the *Caledonian Orogeny* or mountain-building period. In the late Silurian and early Devonian the sediments were intensely folded and in some cases metamorphosed before being uplifted to form the Caledonides, a great mountain range trending across the present-day areas of Scandinavia, Britain and the North Eastern United States. The rocks were tightly folded along SW–NE lines with the greatest deformation being in the Scottish Highlands. There the Moine and Dalradian sediments were metamorphosed and sometimes heaved into great nappe structures. The Moines were also thrust along and up the *Moine Thrust Plane* so that they rode over younger Cambrian sediments (*see* Fig. 65).

The geography was obviously greatly changed by these events with a proto-American area being welded to a proto-European one (Fig. 66). The sea was pushed right away to the south of the Caledonian mountains and the history of the British Isles now took shape in two separate areas (Fig. 67). In the region of the mountains a series of continental deposits were being laid down in the hollows of the land surface—the *Old Red Sandstone* facies (Old to distinguish it from the New Red Sandstone which was formed during later Permo-Triassic times). In the ocean to the south a series of sediments comprising the *marine facies* were accumulating.

The Old Red Sandstones consist largely of conglomerates, red sandstones and marls. By this time the Caledonides of Britain had moved into the arid belt south of the equator and it is thought that these "red-beds" reflect the semi-arid conditions. They probably represent detritus fans and lake deposits although in the South Wales and Shropshire area spreads of shales, marls, and grits have been interpreted as having formed on a great delta plain adjacent to the sea. Similar deposits to these are found along the same belt in Ireland. Further north the more typical continental deposits are found in the Cheviots, the Midland Valley, around Lorne and Glencoe, at Caithness, and in Northern Ireland. They contain little fauna except for occasional fish remains.

The marine facies consists largly of shales or slates, gritstones and limestones with a coral-cephalopod fauna. They outcrop

Fig. 65.—*Simplified structural map of the British Isles.*

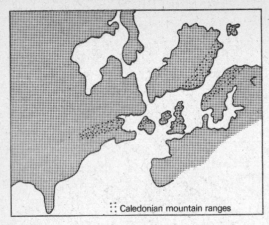

:: Caledonian mountain ranges

FIG. 66.—*Development of the Caledonian mountain ranges.*

mainly in South Devon. In North Devon, on the other side of the Devon synclinorium, marine beds interdigitate with continental ones. The typical marine beds of South Devon have undoubtedly been thrust very much closer to the Caledonian continent than they originally were by post-Carboniferous plate movements.

5. Carboniferous times. The Carboniferous Period, which began about 345 million years ago and lasted for about 65 million years, again saw a dramatic change of conditions. The long period of Devonian erosion had gradually reduced the relief of the Caledonides so that the Lower Carboniferous seas were able to invade many parts of this landmass as they spread in from the south. The Carboniferous strata now outcrop over wide areas. They occur in the South West Peninsula, South Wales, North Wales, the Pennines and the Lake District, the Midland Valley of Scotland and much of Central Ireland.

As might be expected, the sediments over such an area are very variable in thickness and in type. In places they are only a few tens of metres thick whilst elsewhere they reach over 2,000 metres in thickness. This can usually be explained as being due to the variable relief of the area that the sea invaded. Thus inundation and sedimentation occurred in some areas long

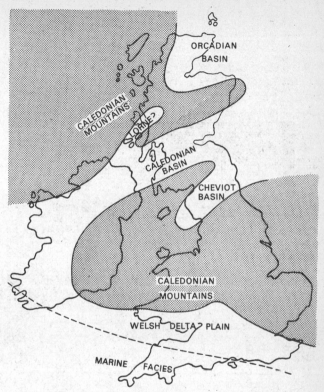

FIG. 67.—*Simplified Devonian palaeogeography.*

before others. Except for the South West Peninsula, where greywackés and shales (Culm) were deposited in a new geosynclinal sea, most of the Lower Carboniferous beds consist of various types of limestone or shale (Fig. 68(*a*)). This *Carboniferous Limestone Series* contains dolomites, oolites, pebble beds and various types of reef structure. The variation in facies has recently been interpreted as being due to a number of phases or cycles of marine transgression and regression. The regressions, which were caused by a general lowering of sea level, left behind

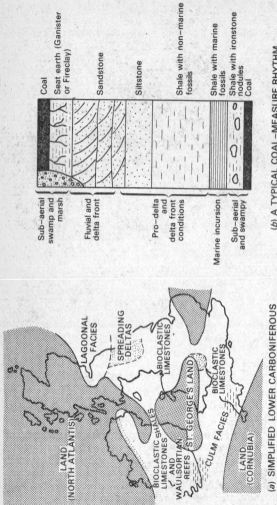

(b) A TYPICAL COAL-MEASURE RHYTHM

Fig. 68.—(a) lower Carboniferous palaeogeography. (b) a typical Coal Measures cyclothem.

some basins of evaporating water in which the dolomitic limestones formed. It must be pointed out here that this area had
by now drifted northwards into equatorial zones, so the high
temperatures necessary for such evaporation would clearly be
occurring.

In the north of England a group of beds known as the *Yore-
dale Series* appear towards the top of the Lower Carboniferous.
They heralded the changes coming in the succeeding Upper
Carboniferous for they consist of limestones, shales, and sandstones occurring in rhythmic sequences and are thought to be
deltaic and pro-deltaic deposits. The beginning of the Upper
Carboniferous saw an increase in the spread of deltaic sands as
the delta area grew southwards from the northern landmass.
These sands together with interbedded shales form the *Millstone Grit Series*. Silting up continued and as a series of tropical
muddy swamps developed so they were colonised by forests and
gave rise to the *Coal Measures*. Rhythmic bedding continued
as a feature of the Millstone Grits and Coal Measures, and may
owe its origin partially to tectonic disturbances (*see* Fig. 68(*b*)).

There is considerable evidence of disturbance during the
Carboniferous period, for not only was there tectonic activity,
but also considerable vulcanicity. Basaltic lavas were poured
out in the Midland Valley of Scotland and some are also found
in Derbyshire and the East Midlands.

6. Permian times. At the end of the Carboniferous and the
beginning of the Permian the disruptions which had punctuated
the Carboniferous period culminated in a mountain-building
episode, the *Hercynian* or *Armorican orogeny*. This occurred as
an African plate closed up to the now combined American–
European plate from the south. As the ocean between them
gradually disappeared so the sediments were crushed and
elevated into a range of mountains stretching a vast distance
across Northern Europe and along the Eastern American
area (Fig. 69). In Britain the strongest effects of this orogeny
are seen today in the South-West Peninsula and South Wales.
There the Devonian and Carboniferous strata have been folded
into a series of east-west trending folds and faults (*see* Fig. 65).
Some of the thrusts suggest movement from the south, and
metamorphism is found in the extreme south of Devon and
Cornwall. Further north in Britain the folding was not so
intense but was nevertheless important. The Pennines were

folded and faulted at this stage and folding and faulting also occurred in Scotland. Undoubtedly many old Caledonian structures were reactivated by these movements. Igneous activity was also widespread. Granites were intruded in the South-West Peninsula with resulting mineralisation. Sills and dykes were pushed into the strata of Northern England, and mineralisation also took place in the Pennines in late Palaeozoic and early Mesozoic times.

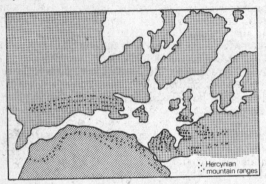

Fig. 69.—*Development of the Hercynian mountain ranges.*

As had happened after the Caledonian orogeny, the geography again underwent a drastic change. The open ocean was pushed well away as this country became part of the great supercontinent of Pangaea (*see* X, **3**). The climate also became increasingly arid as the British area drifted into the subtropics north of the equator. Thus the Permian and Triassic periods are largely marked in this country by the deposition of continental deposits —the *New Red Sandstone*—although further south in Europe marine fossils show that a considerable change in life occurred between the two periods. It is for this reason that they are placed in separate eras, the Permian in the Palaeozoic, and the Trias in the Mesozoic. The British Permian beds were formed in two sorts of environment (Fig. 70). Most are red or yellow breccias, conglomerates, sandstones, and siltstones; they are interpreted as desert basin deposits washed out of the highland regions by occasional rainstorms and then further transported by wind action. Some of the sandstones reveal fossil dune

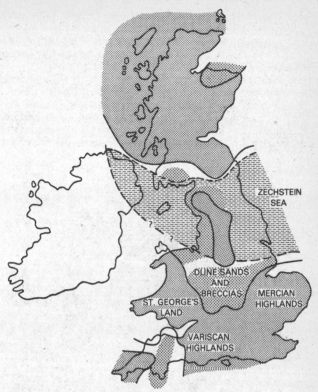

Fig. 70.—*Simplified Permian palaeogeography.*

structures. In the eastern part of the north of England, however, was a saline sea, one of a set of basins that stretched across the present North Sea area to link with the German *Zechstein basin*. It is thought that this was an inland evaporating sea which was periodically recharged as a thick series of dolomitic limestones (the *Magnesian Limestone*) and evaporite salts were laid down.

MESOZOIC ERA

7. Triassic times. The Triassic System is so called because in Germany it was divided into three series: the Bunter, Muschelkalk (a fossiliferous limestone) and Keuper; but in Britain this has never proved a possible subdivision through the absence of the Muschelkalk. Yet the terms Bunter and Keuper have been utilised for the British deposits although there have been many difficulties in correlating them throughout the country, and although they do not altogether equate with the German series of the same name. The climate during Triassic times continued to be warm and arid particularly during the early phases when sediments were essentially the same as those of the Permian. Throughout the Permo-Trias the Hercynian uplands were rapidly being eroded and the basins were gradually filling with

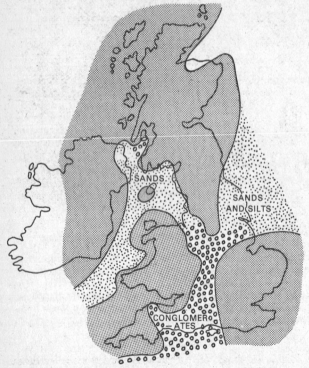

Fig. 71.—*Simplified Triassic palaeogeography.*

sediment. Thus the relief had been very greatly reduced by later Triassic times and the Keuper deposits helped to bury the former Hercynian landscape (Fig. 71). Evidently shallow lakes developed in parts of northern England and the Midlands in which red marls and evaporites accumulated during the Keuper. The salt deposits of Cheshire formed in this way.

8. Jurassic times. During Jurassic times the great continent of Pangaea began to split up and new oceans started to form; and as the British area continued its northern drift so the climate became more humid. The seas once again spread in to cover much of the now flattened landscape, and a variety of shales, sandstones, limestones (often oolitic) and ironstones were deposited.

The outcrops of Jurassic strata are extensive. The main one runs in a great curve from the Dorset coast north-eastwards to the Cleveland Hills of North Yorkshire. Over large areas the beds dip gently to the east or south-east. The more resistant rocks give rise to westward facing escarpments whilst the softer ones form the vales between. The succession thus youngs to the east with the oldest beds outcropping in the west.

The Lower Jurassic deposits are mainly a series of bluish grey shales with some limestones and ironstone beds, (*see* Fig. 72). They are known collectively as the *Lias* and in places yield a rich fauna of ammonites and lamellibranchs. *Oolitic* limestones dominate the Middle Jurassic but in Yorkshire deltaic conditions prevailed as they did also in the Scottish area. Sandstones, siltstones and shales replace the oolitic limestones of further south. The Upper Jurassic strata consist of an alternation of shales, sandy limestones and limestones. The more important shale beds are the *Oxford Clay* and the *Kimmeridge Clay* whilst the *Corallian* forms one of the more prominent limestone members. Two very well known limestones—the *Portland Beds* and the *Purbeck Beds*—occur at the top of the succession but are only present in southernmost England.

The Jurassic seas were relatively shallow and there is evidence that some deeper basins were separated from each other by "shallows" or "swells." Silting up of the seas began in Scotland early in the Jurassic and gradually extended southwards. Thus towards the end of the period only a relatively small area of southern England was subject to marine sedimentation, and

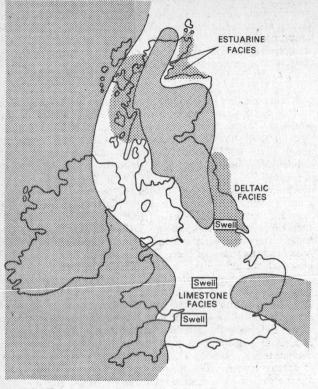

FIG. 72.—*Simplified middle Jurassic palaeogeography.*

even here conditions finally gave way to terrestrial and fresh water as an elevation is thought to have taken place.

9. Cretaceous times. The Cretaceous period which spanned some 71 million years saw the initial splitting which produced the American and European plates as the North Atlantic started to open up. The climate became cooler than in Jurassic times and possibly turned rather arid in the British region.

The Cretaceous rocks form a broad outcrop which extends

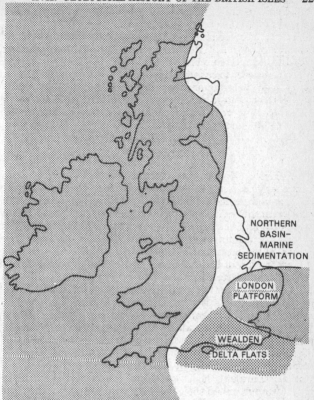

NORTHERN
BASIN-
MARINE
SEDIMENTATION

LONDON
PLATFORM

WEALDEN
DELTA FLATS

Fig. 73.—*Simplified lower Cretaceous palaeogeography.*

from Flamborough Head in Yorkshire to the coast of Dorset.
This same outcrop also extends eastwards to occupy the
Wealden area. Other outcrops occur in Northern Ireland and
the Western Isles of Scotland.

The Lower Cretaceous strata consist chiefly of clays and
sandstones. It is thought that they were initially deposited in
lacustrine conditions in the Weald before the sea spread into
this area of delta mudflats. Further north, in Lincolnshire and

Yorkshire, on the other side of a Palaeozoic massif, marine sedimentation was occurring (Fig. 73).

The Upper Cretaceous produced a marked change as a remarkable transgression of the seas across Britain took place and probably inundated most of the area. Laid down in these Upper Cretaceous seas was the *Chalk*, a remarkably pure and fine grained, white limestone. In places the Chalk is over 300 metres thick and contains very little terrigenous material, thus suggesting very stable conditions and very little erosion of any neighbouring landmasses.

CAINOZOIC ERA

10. Tertiary times. The Tertiary Period saw yet another great change as a major elevation occurred. Most of the British area again became a landmass (probably not of very great relief) with the sea fluctuating in position around South-Eastern England (Fig. 74). The northward drift of the North European area was continuing and the climate which was warm or subtropical at the beginning of the period gradually became cooler. The separation of the American and Eurasian continents also continued and Greenland broke away from Europe. The widespread volcanic activity which took place in North Western Scotland and North Eastern Ireland in Eocene times probably resulted from this opening up of the Atlantic Ocean. Perhaps the most dominant episode of the period, however, occurred in mid-Tertiary times when Italy was driven northward as part of the African plate to cause the building of the Alps. The former great ocean known as Tethys had been closed as the southern continents such as Africa and India moved northwards and now approached their northern counterparts. The ripples of the Alpine orogeny were felt as far north as Britain, and the Mesozoic and Tertiary sediments of South-Eastern England were squeezed into a series of east-west trending folds (*e.g.* Wealden pericline) (*see* Fig. 65). Further north there was renewed movement along old faults, and areas such as the Pennines, the North Yorkshire Moors, and the Lake District were flexed and elevated. It has in fact been estimated that Britain has undergone a regular uplift ever since the main Alpine episode of about 1 metre every 15,250 years.

Tertiary outcrops are found mainly in the London and Hampshire Basins. The strata are largely unconsolidated sands and

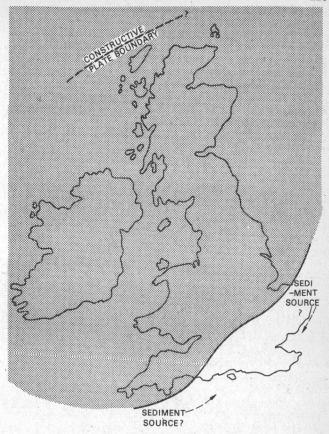

Fig. 74.—*Simplified Tertiary palaeogeography.*

clays. They represent a mixture of marine and fluviatile
sediments deposited where a river system or systems were
draining from western landmasses to the sea in the south-east.
Many of Britain's planation surfaces were also probably cut
late in the Tertiary by river or, in some cases, marine action.

11. Quaternary times. Between about one and two million years ago considerable oscillations in temperature began as the northern hemisphere countries closed in around the Arctic Ocean. The colder phases led to the accumulation of vast quantities of ice over parts of Europe, North America and Siberia. The British Isles were clearly affected by some of these glacial phases although it is not entirely clear how many major glacial advances actually occurred. There is good evidence from many parts of England of at least two glaciations, and some authorities would claim three.

The last of these glaciations has taken up most of the last 100,000 years—in fact only ending about 10,000 years ago. Thus its effects on the landscape remain remarkably fresh.

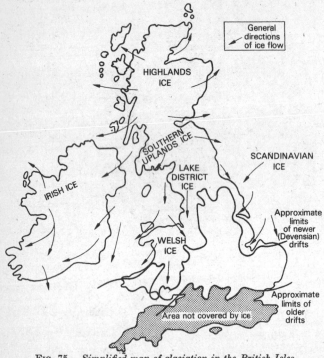

General
directions
of ice flow

HIGHLANDS
ICE

SOUTHERN
UPLANDS ICE

SCANDINAVIAN
ICE

LAKE
DISTRICT
ICE

IRISH ICE

WELSH
ICE

Approximate
limits
of newer
(Devensian)
drifts

Approximate
limits of
older
drifts

Area not covered by ice

Fig. 75.—*Simplified map of glaciation in the British Isles.*

During this glaciation ice caps developed in the highland regions and glaciers flowed out into the surrounding lowlands. Ice also crossed the North Sea from Scandinavia and reached the eastern coast (*see* Fig. 75).

The ice and its associated meltwaters produced considerable erosion especially in upland areas such as the Scottish Highlands, the Lake District, the Northern Pennines and Wales. Valleys were deepened and straightened and numerous cirques were incised into mountain sides producing arêtes and allied features. In the lower lying areas the chief legacy of the ice was moraine such as the boulder clay spreads in the Vale of York, and the hummocky or drumlinoid till found in the Solway Plain or the Craven Lowlands. Thus a great variety of both erosional and depositional landforms comprising moraines, drumlins, eskers, kames, lake beds, glacial spillways, glacial troughs and sub-glacial channels, cirques, arêtes and many others can be seen in the British landscape as a result of the last glaciation.

The oscillations in climate and ice formation caused variations in sea level which, in turn, led to the formation of river terraces and of raised beaches—the latter can be clearly seen around many parts of the coast such as Western Scotland—while the final retreat and disappearance of the ice resulted in a rise of sea level which breached the Straits of Dover and gave the British Isles their present appearance.

PROGRESS TEST 17

1. Describe the Pre-Cambrian rocks of Scotland. What evidence is there to indicate their age relationships? (2)

2. Give a brief description of the geological history of Great Britain during Lower Palaeozoic times. (3)

3. Describe the facies of the Devonian system and suggest how they may be correlated. (4)

4. Explain the formation of the Millstone Grit and Coal Measure series in the British Isles. (5)

5. Outline the evidence which suggests that there were strong earth movements in Great Britain during either the Caledonian or the Hercynian orogenies. (3, 4, 5)

6. Compare and contrast the Old Red Sandstone and New Red Sandstone environments. (4, 6)

7. Give a description of the rock types and fossils found in the British Jurassic and Cretaceous systems. (8, 9)

8. Outline the main geological events that occurred in the British area during Tertiary and Quaternary times. (10, 11)

CHAPTER XVIII

APPLIED GEOLOGY

INTRODUCTION

1. Geology in our everyday lives. The previous chapters in this HANDBOOK have been concerned with the basic facts, principles and concepts of the science of geology. An attempt has been made, if only briefly, to describe the interior and exterior of the earth and to indicate the methods by which the facts of geology have been obtained. An attempt has also been made to show how ideas and theories, which attempt to explain certain geological features, are related to the geological facts as we know them. The geologist does not pretend to have found out everything about the geological past: on the contrary, there are many things that continue to puzzle him.

Having described in a fairly simple and elementary fashion what is known about the earth and the processes which have helped to shape the physical landscape, it will be useful to conclude with a short account of the way in which geology is of significance in our everyday lives. The applications of geology to everyday life form *applied geology*. Since these usually have important economic implications, applied geology is sometimes called *economic geology*. The two are not, however, entirely synonymous but for present purposes we can think of the economics of geological materials as being applied geology.

2. Application of geology. Nowadays geological investigations are a necessary preliminary to all operations involving mining, whether it be for coal, metallic ores or chemical deposits, boring for water, petroleum or natural gas, or engineering projects such as tunnelling, bridge-building, the construction of dams or the erection of large buildings. Unless the geologist is consulted and his recommendations duly taken into consideration, wasted time, energy and capital may—indeed is likely to—result and disaster may follow.

Man's primary needs are food, water and shelter but the

civilisation he has built up is fundamentally based upon the natural resources which Nature has offered him—supplies of energy and raw materials—together with his scientific knowledge and technological skill. A large part of the natural resources available to man—water, sources of energy, metals, building materials, chemicals—have a geological base.

It is impossible within the scope of a short chapter to do more than describe in the briefest way some of the numerous apsects of applied geology; we have had to be selective and attention is drawn to four important aspects:

(a) The prospecting for, and mining of, ores.
(b) The occurrence and exploration of oil and natural gas.
(c) The provision of water supplies.
(d) Civil and structural engineering problems.

ORES: PROSPECTING AND MINING

3. Ores. An ore is a rock containing a sufficient proportion of a metal to make its extraction an economic proposition. For an ore to be mined, it must be economically profitable to do so. However, it is important to note that a metalliferous rock which is valueless as a source of a metal at one time may be worth while mining at another. An ore's changing economic worth is mainly related to two factors:

(a) the value of the metal: for instance, if the price of the metal goes up it may be economically worth while to mine inferior ores;
(b) improved techniques of mining or improved extractive processes may render a hitherto uneconomic ore economic.

In the case of common metals, such as iron, the yield of metal from the ore should be of the order of 30–60 per cent, although some ores, such as taconite, which undergo sintering, may have less than 30 per cent of metal in them. With rare or precious metals the yield may be under 1 per cent.

Some minerals are very common and very abundant constituents of the earth's crust, e.g. iron and aluminium, but in the case of the latter its extraction and use was dependent upon the development of electric power for separating the metal from its ore. Titanium has been described as a "wonder metal" but its future use will depend upon the ease and cheapness by which it can be obtained.

4. The occurrence of minerals. Although mineral resources are widely spread throughout the crust, comparatively few are found in sufficiently concentrated quantities to justify their commercial exploitation and, for that reason, the mining of minerals is usually confined to localised areas.

Minerals occur in three ways depending upon the geological conditions under which they were formed:

(a) *In lodes or veins.* When igneous rocks were intruded into the crust various liquids and gases found their way into cracks and fissures, eventually cooling and congealing to form lodes and veins of metalliferous minerals and gangue minerals. A number of these metals often occur in association with each other, *e.g.* copper with nickel, silver with lead and zinc, iron with manganese. Diamonds occur in pipes of igneous rock.

(b) *In sedimentary beds.* Some minerals are laid down in horizontal sheets or layers; for example, many iron ores, such as the Jurassic iron ores of Eastern England, and bauxite, the ore of aluminium. Iron and aluminium are the most notable of the so-called "bedded ores." Evaporites, such as halite, gypsum, and related minerals which have been formed by evaporation and precipitation also occur in beds.

(c) *In alluvial deposits.* Many minerals are found in alluvial or gravel deposits at the bases of hills or in valley bottoms. Usually originating in the veins and lodes of nearby hills, the minerals, exposed by erosion, have been washed downslope and dumped in the debris accumulating at lower levels, *e.g.* the tin of Malaya, the "placer" gold of Yukon in Canada.

5. Distribution of mineralised regions. Whereas coal is chiefly associated with Carboniferous basins, and petroleum and natural gas with fold structures in zones of sedimentary rocks, metallic minerals are usually (though by no means exclusively) found either in ancient "shield" areas and old, stable plateau masses which have often suffered metamorphism or in regions of young fold mountains which have experienced more geologically recent earth movements and intrusions. The Baltic and Canadian Shields and the Plateaus of Brazil, Africa, India and Australia provide examples of the former, the Andes and the Rockies examples of the latter. Mineralisation is found chiefly at destructive plate boundaries.

Some metals or ores are closely associated with certain rock formations or conditions. For example, the Carboniferous Limestone in England is peculiarly associated with lead zinc and fluorite; iron ores, such as limonite, are found in Jurassic beds; bauxite is found more especially in tropical regions where the great heat and abundant moisture has caused deep weathering.

6. Geological prospecting. In the old days prospecting was carried out very much "by God and by guess." Today, the mining prospector is a highly qualified and carefully trained person—usually a geologist who is an expert in mineralogy, petrology and structural geology. But, in addition to his expertise as a geologist, he commonly calls in the help of others who are specialists in the fields of geophysics, geochemistry and geobotany. In fact these days initial geological prospecting is usually done by geophysical and geochemical methods rather than by the more traditional methods of the geologist. For example, the rich Boliden deposits of Norrland in Sweden, which contain a varied assortment of minerals, including gold, silver, lead, zinc, copper and arsenic, were found by geophysical methods. Edmunds quotes the case of an important gold find in South Africa which was located by geophysical means in an area which geological deduction had indicated would be likely to be unyielding. (*Geology and Ourselves*, Hutchinson, 1955, p. 199.) The use of geochemistry to detect trace elements in the soil is often applied nowadays as is the use of plants to locate ores. In more recent times vegetation has been much used to locate zones of mineralisation; the occurrence of specific plant species has led to the location of areas of copper mineralisation in Zambia and Brazil, to lead and zinc mineralisation in Australia, and to iron-ore fields in Australia and Venezuela.

7. The exploitation of mineral resources. The utilisation of mineral resources is dependent upon a variety of factors:

(*a*) *The size of the deposit and the percentage of metal content in the ore.* Ores are often termed high-grade or low-grade according to the abundance of a particular metal in the ore, or the occurrence of usefully associated metals.

(*b*) *The depth at which the ore occurs.* If it is near the surface it may be worked by open-cast methods. Deep mines are costly to establish and work.

(c) *The distance of the ore body from the consuming market* and the transport facilities available for the shipment of the ore.

(d) *The existence of a satisfactory labour supply* which may be difficult to recruit in remote, inaccessible and unattractive areas.

(e) *The availability of capital* which may be required in large quantities for large-scale enterprises involving expensive mining plant and perhaps transport facilities.

(f) *The presence or absence of fuel and power resources* required in the processing of the ore, *e.g.* for concentration, smelting and perhaps final refining.

(g) *The international demand for the metal.* If the demand is great enough, mining will often be undertaken in the most unattractive environments, *e.g.* Port Radium (for uranium) in arctic Canada.

MINERAL OIL AND NATURAL GAS

8. The origin of oil and gas. Having considered minerals in general we can now turn our attention to a particular one. Mineral oil, from which many different kinds of oil are eventually derived, is generally thought to be due to the decomposition of tiny marine creatures, minute plants and animals similar to the plankton which occurs in the oceans at the present time, which sank to the bottom of lagoons and seas and became intermingled with the muds on the lake and sea floors. Ultimately this dead organic matter became entombed and preserved in the sediments which were deeply buried, often to depths of several thousand feet.

We are not very certain how this organic matter imprisoned in the rocks came to be transformed into oil. It may well be that petroleum has originated in several ways. Formerly, it was thought that great heat and pressure were necessary for the formation of oil but this old idea is tending to be discarded and modern research workers are suggesting the quick burial and quick transformation into oil without the need for great pressure exerted by the tremendous weight of overlying rock strata. This new idea, however, does not rule out the earlier one. As we have just said, several modes of origin may be involved in the formation of oil. The decomposition of the organic matter by bacteriological action, however, appears to be a necessary stage in the process.

9. The occurrence of oil and gas. During the time in which the dead organic matter was being entombed and transformed into oil, sediments continued to settle and accumulate on the lagoon or sea floors. As a result of the great loads of sediment, the deeper layers were squeezed, compacted, hardened and turned into rock. Under such great pressure the muds in which the organic material was originally laid down became dense and impervious to fluids. In the sandstones, however, which were made up of larger mineral grains, pore spaces remained and in these tiny spaces water particles (for the sands had originally been laid down in sea water) continued to be present. The minute bubbles of oil and gas, products of the alteration of the organic matter buried with the mud at the time it was originally laid down on the sea bed, were squeezed out of the muds as they were being compacted and moved into the adjacent sandstones whose pore spaces were able to accommodate them. Due to the gradually increasing pressure, the particles of oil and gas, as well as the sea water, migrated upwards until their movement was halted by an overlying layer of impervious mud.

The presence of an impervious "ceiling", *i.e.* cap rock, meant that the oil and gas could now move only horizontally and so they tended to spread laterally along the sandstone stratum, floating on top of the water which permeated the pore space in the rock. As this lateral movement took place the minute oil particles gradually coalesced to form larger and larger bubbles. In due course, the oil bubbles accumulated in some dome-shaped concavity or irregularity in the strata where they lay trapped. In such "traps" they were confined by the hydraulic pressure of the brine on which they floated and it is these trapped accumulations of oil and gas that we call oil pools (*see* Fig. 76).

10. Exploration for oil. Some kind of geological "trap" is necessary for oil to be present. Many kinds of trap occur in the rocks but the simplest, and a very common form of trap, is an anticline where the strata have been gently arched upwards. In a trap of this kind, oil often floats on top of salt water while above the oil there is frequently natural gas. It should be noted that oil may occur without gas and that gas may be present without oil.

Wherever there are areas of sedimentary rocks which have not been violently folded there is always the possibility of finding oil. The oil prospectors of half a century ago experienced no great difficulty in locating deposits of oil; they merely searched for areas where oil oozed out at the surface or where natural gas escaped, bored their holes and waited for the greenish-black petroleum to gush out. Early prospecting, however, was very much a matter of hit and miss but since the demand was

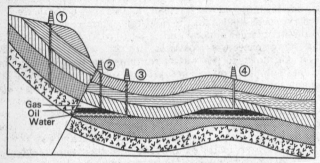

FIG. 76.—*Cross section showing oil-traps.* The diagram shows two kinds of "traps" in the rock strata of the earth's crust in which oil may be found. On the right, the oil is trapped in an upfold or anticline; on the left, a fault in the rocks has led to a movement in the rock layers and an impervious layer has formed a barrier which dams up the oil. Well 1 hits a dry area; well 2 taps oil; well 3 draws water; and well 4 taps natural gas.

small sufficient oil could be found by such unscientific methods. The old-fashioned practice of "wildcatting," as the random drilling was called, has now been replaced by highly scientific methods of oil exploration. Where successful oil wells had been drilled, geologists carefully studied the rock structures and so were able to work out where the oil was likely to be found. Petroleum geologists are still not able to do this with complete accuracy. For example, in 1947 a rich oil field was discovered in the prairie province of Alberta, in Canada, but 30 years had been spent in finding it and 134 wells having a total depth of 160 miles had been dug during the quest !

In an attempt to locate oil deposits, geologists use many geophysical techniques (*e.g.* seismic, gravimetric and magnetic surveys) which can give an indication of whether oil is likely to be found in a given place. In the case of seismic surveys the geologists create artificial "earthquakes" by firing explosives: the "shocks" can be measured by instruments (similar to the seismometers which measure earthquake waves) and the readings help to indicate the arrangement, density, permeability, etc., of the rocks which, in turn, indicate whether oil is likely to be found by boring.

HYDROGEOLOGY

11. Importance of water. In no part of our everyday life is geology of more direct importance than in the provision of adequate water supplies. Though so common, water is one of the most precious of all the natural resources. Water is indispensable for life but, quite apart from its crucial importance in this respect, it has numerous other uses, for example:

(*a*) *domestic uses*—drinking, cooking, bathing, washing, sanitation, garden watering;

(*b*) *municipal uses*—fire fighting, street cleansing, in hospitals, in schools, swimming baths;

(*c*) *agriculture*—watering stock and growing crops, especially in arid and semi-arid regions where irrigation must be practised;

(*d*) *industry*—for cleaning and processing commodities, for steam generation, for cooling purposes;

(*e*) *power*—in earlier times running water turned water wheels but nowadays the water is mostly used in the production of hydro-electric power;

(*f*) *effluent disposal*—water is used to transport wastes from urban and industrial areas.

(*g*) *transport*—waterways (rivers or canals) are utilised for the cheap carriage of, in particular, bulky and heavy commodities;

(*h*) *food supplies*—water areas provide habitats for fish which in some areas may be significant as an article in the diet;

(*i*) *recreation*—water may have an amenity value, providing opportunities for sporting activities or merely serving an aesthetic purpose.

12. The demand for water. Since water has multiple uses, the demand for it is great and growing. The progressive demand for increased supplies of water in most parts of the world arises from three main causes:

(a) The gradual rise in individual consumption consequent upon higher standards of living.

(b) The global increase in population—in excess of 50 millions a year.

(c) the insatiable demands of industry, a demand which is closely linked to the two fore-going factors.

In providing water supplies, the water engineer is concerned with the *quantity* and the *quality* of water. For most purposes the provision of sufficient supplies of water is not enough in itself: the water must also be of considerable purity and freshness, and for certain industrial uses it must be free from particular chemical constituents.

Some indication of the colossal demand for water in our present society is provided by the fact that in Britain approximately 1,500 million gallons (6,820 million litres) are consumed daily!

13. The role of geology in water supply. Firstly, to ensure that ample and constant supplies are available it is necessary to construct reservoirs to collect and hold the run-off. Reservoirs cannot be sited anywhere; their location is, in fact, closely determined by a variety of geological factors, *e.g.* preferably in capacious valleys which can be relatively easily sealed off by a dam, in areas of impermeable rock so that the water does not percolate and seep away, in areas free from geological faults which might permit seepage or make dam construction difficult or impossible.

Secondly, underground supplies of water are closely related to the character and disposition of the rocks. The porosity and permeability of rocks determine their capacity for holding water; the ability of rock to hold water depends upon the aggregate volume of the intercommunicating voids (pores and fissures) within the rock. Some rocks possess a great capacity for containing water, *e.g.* chalk, limestone and many sandstones, therefore the location of such rocks is important in the provision of subsurface supplies.

Thirdly, the sub-surface structure is an important factor in influencing the occurrence or otherwise of underground water. For example, the presence of an artesian well is entirely governed by geological factors.

Fourthly, the chemical character of the rocks is likely to influence the quality of the water. We are all familiar with the terms "soft" and "hard" water. Water becomes "hard" because of the presence in solution of calcium, magnesium and iron compounds; such water does not readily form a lather with soap, hence it is of no use in the textile industry which needs "soft" water. Hardness in water may be:

(a) *temporary*, *i.e.* when it contains soluble bicarbonates;
(b) *permanent*, *i.e.* when it contains sulphates.

These are only some of the ways in which the geological conditions may influence the occurrence, quantity and quality of water.

CIVIL AND STRUCTURAL ENGINEERING

14. Geology in relation to engineering. F. H. Edmunds, in his interesting book on the importance of geology in our everyday lives (*Geology And Ourselves*, Hutchinson, 1955, p. 156), has written:

"There is a close relationship between foundation works for buildings, bridges, dams, and other structures, and the strata that support them. All these structures rest *in* the strata, rather than on the ground surface; consequently foundations and supporting beds have to be welded into a whole, as it were. Geological factors thus enter into the initial stages of all civil and structural engineering work. This applies in principle whatever the size or style of erection the foundations have to carry, or whatever the scale of excavation; whether for minor works such as house building or laying sewerage, gas or water mains, or for major activities connected with the construction of marine docks, railway tunnels and other large works."

Geological considerations are not always taken into account, nor need they necessarily be, for the use of a little common sense is often all that is required. But, for any major undertaking, it would be the height of folly to ignore geological factors.

15. Requirements of a site investigation. If we may be allowed to quote Edmunds again, he says that any site investigation at the present time should include, amongst other things, an examination of the following geological points:

> "(a) Nature of the strata; whether hard rocks or soils: if the latter, whether cohesive or non-cohesive; their composition; moisture content; grading and compactness; and their behaviour under specific laboratory tests.
>
> (b) Characteristics of the mass of each type of rock, e.g. whether bedded, or not; if bedded, thickness and variations in the thickness of the beds; the nature of fissures and joints in hard rocks.
>
> (c) Arrangement of strata; alternations of different types; whether strata are horizontal, more or less uniformly inclined or faulted; whether broken or shattered.
>
> (d) The arrangement of the strata and their reactions to the retention and circulation of water.
>
> (e) The level of the water-table.
>
> (f) Whether sulphates and other mineral salts are present." (Op. cit., p. 160.)

16. The problem of water. The presence of water in the soil and rocks is of special significance to the structural engineer, for it may create problems, which in turn may increase construction costs in engineering work. The need to effect adequate drainage is usually a pre-requisite to any building construction. Unless water is effectively drained away land slips and associated phenomena may occur. Mining operations below ground are frequently troubled by underground water and the mechanical pump was one of the first contrivances to be developed during the early days of the Industrial Revolution when coal began to be mined in quantity. The driving, and subsequent maintenance, of tunnels is closely associated with water difficulties.

Conversely, loss of water from the ground may create difficulties. For example, in the Fenlands of Eastern England, drainage operations over a long period have led to the shrinkage of the peat which, in turn, has resulted in the lowering of the land surface. Pumping of water from the ground may lead to differential subsidence and structures built in such areas may become cracked, tilted or even collapse.

17. Dam construction. The building of dams to impound surface water provides a good example of geology in relation to engineering. In collecting and storing the run-off fraction of precipitation, the basic problem facing the civil engineer is the choice of site. His choice will be largely determined by two considerations: topographical and geological. The topographical consideration will lead him to choose, if it is at all possible, a long, deep, narrow valley which can be easily sealed off by a short dam. The geological consideration involves the choice of a site which is structurally capable of supporting the dam; a dam is a very weighty structure and the rocks of the valley floor must be able to support the dam wall, moreover the valley sides must be able to sustain the great pressures of the impounded water.

A particular valley may be, from the topographical point of view, very promising, but the geological conditions may cause it to be more or less useless for water impoundment. Water-tightness is very important but it may be seriously impaired by such geological conditions as porous rocks or the presence of faults. There have been cases in recent times of serious mishaps where dams have collapsed, *e.g.* the one at Fréjus, in the French Riviera, where either the dam structure was faulty or insufficient care had been taken in the choice of the site.

PROGRESS TEST 18

1. What is applied geology? How important is geology in our everyday lives? (1)

2. Describe the different ways in which minerals may occur. (4)

3. In former times geological prospecting was done largely "by God and by guess." How does modern geological prospecting differ? (6)

4. What factors influence the exploitation of mineral resources? (7)

5. Describe (a) the origin, and (b) the occurrence of petroleum and natural gas. (8, 9)

6. Write an essay on water. (11, 12, 13)

7. In what different ways is water of importance to man, his life and his activities? (11, 12)

8. Show how the demand for water has grown. (12)

9. How do geological conditions affect water and water supplies? (13)

10. What are the main requirements of a site investigation? (15)

11. In what ways may the presence of water create problems for the engineer? (16)

12. Briefly describe the meaning of the following: hydrogeology, lodes, oil-trap, geochemistry, hard water, bedded ores. (4, 6, 9, 13)

CHAPTER XIX

GEOLOGICAL MAPS

INTRODUCTION

One of the most important areas of work for the geologist is that of map-making. Geological maps are made from field observations of rock outcrops, topography, soils, and vegetation. Since continuous outcrops of rock are seldom encountered except in mountainous regions some of this mapping may be interpretative as well as observational. Thus considerable training and experience is required for the mapping of complex geological terrains. Yet, as well as making maps, it is also part of a geologist's training to learn how to read and interpret geological maps. Maps show the patterns of strata in two dimensions only, but the geologist has to think of them in three dimensions, for it is essential that the subsurface structure be known. Thus much of map interpretation is aimed at developing methods of producing three dimensional models from two dimensional maps. Whilst it is beyond the scope of this book to develop a full course in geological map interpretation, it is hoped that the following sections will form a useful introductory guide.

GEOLOGICAL MAPS

1. Types of geological map. Geological maps are produced at various scales and may show different types of information. For example, some are produced essentially as tectonic maps and concentrate upon faults and folds and their age of formation. Others are designed for economic purposes and give specialised information on mineral deposits, for example, or hydrogeology. Geophysical maps are also produced and may show the results of gravity or magnetic surveys. Maps of all these kinds are made and published in Britain by the Institute of Geological Sciences. Most of their maps, however, show the general geology of an area and display the following information:

(a) The distribution of the strata and the position of the boundaries between them.

(b) The age of the rocks shown; the key usually gives the stratigraphical position and name of the rock.

(c) The thickness of beds; this can be read from a scaled vertical column of strata shown in the map margin.

(d) Various structural elements; features such as the dips of the strata, the position of faults and their throw may be shown on the map. The stratigraphical position of unconformities is also often indicated on the margin.

(e) Information is given on some maps of particular fossil bands and localities.

(f) Some features of economic importance; the outcrops of mineral veins or coal seams, for example.

(g) A geological cross-section of the area shown on the map.

(h) On a few maps, the $2\frac{1}{2}$ inch to 1 mile, (1 :25,000 series) a very brief description of the geology of the area is given on the sheet margin.

(i) The one inch to one mile (1 :63,360 series) and the 1 :50,000 series are sometimes published in separate *solid* and *drift* editions. The former show only the pre-Pleistocene beds and omit any surface covering of boulder clay or peat, etc. The latter show all superficial deposits, but the positions of the solid strata hidden beneath them are sometimes indicated by a series of lines.

Many of the geological sheets produced by the Institute of Geological Sciences are complex in detail and appearance. Thus it is both better and usual for the beginner to learn the elements of map interpretation from simplified geological sketch maps which are constructed to illustrate specific problems. Nevertheless, it is also very helpful and important for the student to familiarise himself with the data given on real geological sheets. It may prove necessary, for example, to abstract information from a particular sheet in order to help with field studies or stratigraphical studies. For this reason a short list of maps is given below. They have been selected to illustrate the various geological features indicated.

1. *Gently dipping strata:* Aylesbury Sheet (Sheet No. 238). A gently south-easterly dipping sequence of Cretaceous strata with a few cappings of Eocene.
2. *Folded strata:* Isle of Wight Special Sheet. A synclinal fold crossing the Oligocene outcrop in the north with a

strong monoclinal feature in the Chalk of the centre of the island.

3. *Folded and faulted strata:* Pembroke Sheet (Sheet No. 245). A series of very tight, largely east-west trending folds in Palaeozoic strata. Reversed faults and thrusts are parallel to the strike and small transcurrent faults cut across and displace the folds.

4. *Unconformities:* Frome Sheet (Sheet No. 281). A rather complicated-looking outcrop pattern but two major unconformities are seen—one where the Trias overlies the folded Carboniferous, and the other where the Cretaceous cuts across the Upper Jurassic strata.

5. *Igneous rocks:* Arran Special Sheet (Solid edition). A varity of intrusions of various ages appear. A granite boss dominates the north of the island and there is a volcanic complex in the centre. Numerous sills occur particularly in the south and a NW–SE trending swarm of dykes are found throughout the island (Fig. 77).

6. *Drift and superficial deposits:* York Sheet (Sheet No. 63). A number of glacial and fluvioglacial deposits are shown together with some recent sediments.

All of these sheets are at a scale of one inch to one mile, although they are slowly being replaced by a 1:50,000 metric series.

2. Interpreting geological maps. The interpretation of geological maps involves two major objectives:

(*a*) An understanding of the structure shown on the map; it should be possible to interpret the rock relationships shown and visualise their arrangement in three dimensions.

(*b*) The construction of a geological history (*i.e.* a chronological sequence of events) of the area from the evidence shown on the map.

The necessary information to fulfil these objectives is again usually obtained in two ways—firstly by visual inspection and secondly by the adoption of various geometrical techniques of analysis.

3. Inspection of the outcrop patterns. This alone can yield a great deal of information. The following may provide some useful guidelines:

(a) *Bedding plane dips:* on some maps the directions and amounts of dip of the strata are shown by dip arrows. If no dip arrows are present but the relative ages of the beds are known then the dip will be in the same general direction as the beds are younging (Fig. 78(a)). One other way in which an idea may be obtained is by examining the outcrop pattern in relationship to the topographic contours:

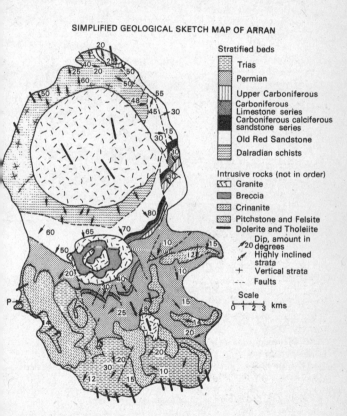

SIMPLIFIED GEOLOGICAL SKETCH MAP OF ARRAN

Stratified beds

Trias

Permian

Upper Carboniferous

Carboniferous
Limestone series

Carboniferous calciferous
sandstone series

Old Red Sandstone

Dalradian schists

Intrusive rocks (not in order)

Granite

Breccia

Crinanite

Pitchstone and Felsite

Dolerite and Tholeiite

⤢20 Dip, amount in degrees

Highly inclined strata

+ Vertical strata

--- Faults

Scale
0 1 2 3 kms

FIG. 77.—*Simplified geological map of the Isle of Arran.*

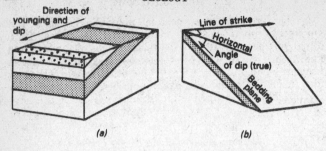

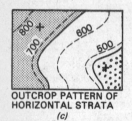

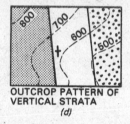

FIG. 78.—*Features of sedimentary outcrop patterns.*

(*i*) Horizontal beds always form outcrops parallel to the contours (*see* Fig. 78 (*c*)).

(*ii*) Vertical beds always form straight line outcrops irrespective of the surface topography (*see* Fig. 78 (*d*)).

.(*iii*) Sloping beds are best observed where their outcrops cross valleys. A bed which is dipping in an upstream direction makes a V-shaped outcrop. The V points upstream but appears much blunter than the V's made by the contours (*see* Fig. 78 (*e*)).

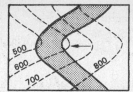

Outcrop pattern of bed dipping
downstream at a steeper angle
than stream gradient

(a)

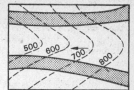

Outcrop pattern of a bed dipping
downstream at the same angle as
the stream gradient

(b)

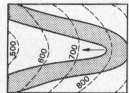

Outcrop pattern of a bed dipping
downstream less steeply than
the stream gradient

(c)

Sequence of intrusion

1. Boss
2. Dyke
3. Sill

Dyke

Boss

SILL

IGNEOUS INTRUSIONS

(d)

FIG. 79.—*Features of sedimentary and igneous outcrop patterns.*

(iv) A bed dipping in a downstream direction but whose dip is
less than the valley gradient also V's upstream, but in this case
the V made by the outcrop is much sharper than that made by the
contours (see Fig. 79 (c)).

(v) A bed dipping in a downstream direction with a slope greater than the valley gradient V's downstream (*see* Fig. 79 (a)).

(vi) A bed dipping downstream at the same angle as the valley gradient forms parallel outcrop traces on either side of the valley (*see* Fig. 79 (b)).

To summarise these relationships, it can be said that the outcrop V points in the direction in which the stratum underlies the stream.

(b) *The relative ages of strata:* if the ages of the strata are not given but the dip is known, then the strata will normally be younging in the direction of dip (*see* Fig. 78(a)).

(c) *The relative thicknesses of strata:* In a uniformly dipping sequence of strata some guide to their relative thicknesses can be obtained from their outcrop pattern. Generally the wider the outcrop, the thicker the bed. Great care must be exercised in reaching such conclusions, however, for the width of outcrop also varies with factors such as the dip of the bedding and the topographical slope.

(d) *Folding:* the existence of folds can be detected by the repetition of beds as a traverse is made across the strike. Once again, however, very careful examination is necessary since horizontal or gently sloping strata in a rugged terrain can give very similar repetitious outcrop patterns (*see* Figs. 80(a) and (b)).

The elements of symmetry of the fold can often be ascertained fairly quickly. An anticlinal fold has the oldest beds exposed at the centre, along the fold axis (*see* Fig. 80(a)). A synclinal fold has the youngest beds at the centre (*see* Fig. 80(c)). The relative dips of the two limbs can sometimes be assessed from the width of outcrop of one particular stratum on the limbs. Assuming that the topography is similar on each side, the limb on which the stratum makes the widest outcrop is the more gently dipping one (*see* Fig. 80(c)).

If a fold is plunging, its limbs gradually close together to form a V-shaped outcrop. If it is an anticline the limbs close together in the direction of the plunge; if it is a syncline they close away from the direction of plunge (*see* Figs. 80(d) and (e)).

It may not be possible to determine the precise age of the folding. Clearly it must be younger than the beds that have been folded and it will also be older than any beds which have not been affected.

(e) *Faulting:* a certain amount of information can be

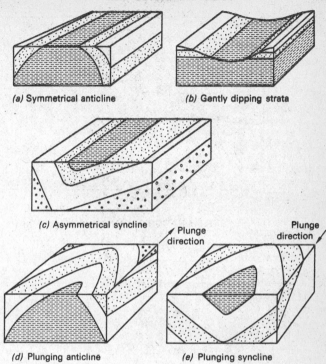

(a) Symmetrical anticline   *(b)* Gently dipping strata

(c) Asymmetrical syncline

(d) Plunging anticline   *(e)* Plunging syncline

FIG. 80.—*Patterns of outcrops produced by folds.*

gleaned by examining the outcrop of a fault on a map. On some geological maps the downthrow side of the fault is indicated by a short tick on the fault line and the actual amount of downthrow is shown alongside. If the downthrow is not marked, it can be quickly determined. Where different beds are brought against each other on opposite sides of a fault line, the younger bed is always on the downthrow side (*see* Fig. 81(*a*)). Care should be taken to check if it is apparent or real. For instance, in dipping strata a vertical downthrow may produce an apparent lateral displacement of beds, and similarly lateral displacement may produce an apparent downthrow. Should the fault show rapid apparent variations

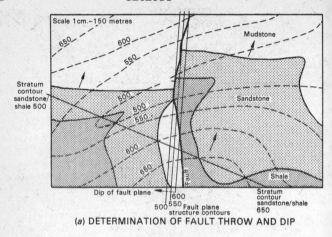

(a) DETERMINATION OF FAULT THROW AND DIP

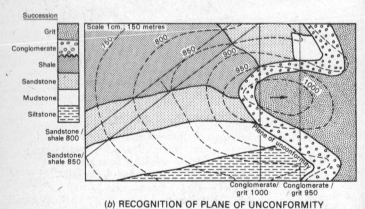

(b) RECOGNITION OF PLANE OF UNCONFORMITY

Fig. 81.—*Determination of faults and unconformities.*

in downthrow it may be that it is a transcurrent fault and the movement has been lateral. Lateral movement can be proved if the fault has moved vertical planes—if it has displaced a dyke or a vertical fold axial plane, for example.

The dip of a fault plane cannot be accurately gauged. However, a general guide is given by its outcrop. The straighter the line of the fault the nearer the fault plane is to the vertical. The more sinuous the fault outcrop the more likely is the fault to be a low angle one.

The age of faulting is determined in the same way as folding. Faults must be younger than strata they affect and older than groups of strata which are not affected.

(*f*) *Unconformities:* oversteps and overlaps can be recognised where the bedding plane of a particular stratum cuts across or truncates other strata (*see* Fig. 81(b)). Care must be taken, however, not to confuse thrust planes with planes of unconformity, for they sometimes have a very similar appearance on a map. Yet thrusts push up older strata above younger, whereas unconformities maintain the normal sequence. An indication of the extent of erosion that occurred prior to the deposition of the unconformable sediments can be obtained by noting the differences of age and the thickness of the beds overstepped.

(*g*) *Igneous forms:* extrusive igneous rocks usually appear on a map as part of a bedded sequence. The same is true of some concordant igneous intrusions (*e.g.* sills), with the result that they are difficult to tell apart. As they are traced laterally they may be seen to thicken or thin or perhaps become slightly transgressive.

Intrusive discordant forms can usually be recognised from the shape of their outcrop. Dykes produce linear outcrops; stocks, bosses and volcanic necks have a roughly circular pattern (*see* Fig. 79(*d*)); very large intrusions such as batholiths are usually elongate features which can be discordant or concordant to the structure.

It should be possible to determine the relative ages of many intrusions by noting the ages of the strata into which they are introduced. An intrusion, of course, must be later than the rocks it is invading but will be older than any rocks which cut across it (*see* Fig. 79(*d*)).

4. Geometrical techniques of analysis. These can vary from simple to very complicated. The simplest and most useful device for the beginner is the construction of *stratum contours* or *strike lines*. These are contour lines drawn on the bedding planes of the strata. They reveal the slopes and shape of a bedding plane in the same way that topographical contours reveal the slopes and shape of the land surface. In order to construct stratum contours the following procedure should be adopted:

(a) Note the positions where a particular bedding plane outcrop crosses a contour of one specific value more than once, if possible at least three times (*e.g.* 700 m contour in Fig. 82(*a*)).

(b) A line should be constructed joining these points and produced to the edge of the map. On simple maps this is usually a straight line.

(c) The line so constructed is a stratum contour and it should be labelled accurately at the map margin. The label should consist of the name of the bedding plane it has been drawn upon, together with the contour altitude. A stratum contour drawn in this way should not be extended across faults and fold axes.

(d) By following the procedure above, a number of stratum contours should be constructed. Should there be insufficient suitable contour-bedding plane intersections it may be permissible to use just one such intersection and construct the line parallel to the original stratum contour. This is usually a satisfactory procedure when beds appear to be in a normal conformable sequence.

Stratum contours can also be constructed from borehole data. If, for example, the *altitude* (not depth) of a particular bedding plane is calculated at three locations (such as in boreholes) it becomes possible to determine the strike direction. The three points are joined to form a triangle, each side of which is then split into convenient divisions based on altitude. These altitudes can easily be calculated by determining the total amount of fall from one end of the line to the other. Two points in the triangle which have the same altitude value are then selected and joined by a straight line. This straight line is the master stratum contour (*see* Fig. 82(*b*)).

5. The significance of stratum contours. When the stratum contours have been constructed and labelled, they can then be used to obtain various sorts of information. Among these are the following:

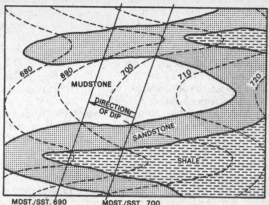

MDST./SST. 690   MDST./SST. 700

Bedding plane drops 10 metres in a horizontal distance of 50 metres(i.e. has a gradient of 1 in 5)

HORIZONTAL SCALE

0 ————————— 50 metres

——— 680 ——— CONTOUR LINE (ALTITUDE VALUES IN METRES)

(a) THE DETERMINATION OF DIP BY STRATUM CONTOUR CONSTRUCTION

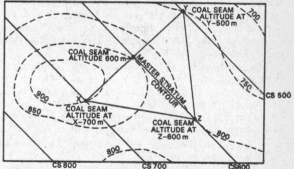

The top of a coal seam is encountered at a *depth* of 200 metres in boreholes X and Z and 250 metres in borehole Y. The method of contruction of stratum contours for the coal seam from this information is shown above.

HORIZONTAL SCALE

0 ———— 400m

—— 850 ——— CONTOUR LINE (ALTITUDE VALUES IN METRES)

(b) THE CONSTRUCTION OF STRATUM CONTOURS FROM BOREHOLE DATA

FIG. 82.—*Stratum contour construction.*

(a) *The direction of dip and its gradient:* the direction of dip is always perpendicular to the stratum contours. Its slope can be seen from the way in which the stratum contours on *one* particular bedding plane drop in altitude (*see* Fig. 82(a)). The gradient of the slope (the dip) is calculated by measuring the perpendicular between two stratum contours (*e.g.* 50 m). The drop in altitude of the bedding plane over that distance can then be read from the stratum contour labels (*e.g.* 10 m) and the gradient calculated (*e.g.* 1 in 5).

(b) *Fold plunge directions and amounts:* where stratum contours cross the axis of a plunging fold they make a V-shaped bend. An analogy can be made with a river valley. Where topographic contours cross a sloping valley they make a V-shaped pattern. In the same way stratum contours drawn on a plunging syncline make a similar pattern. For an anticline the stratum contours close in the direction of plunge; for a syncline they open in the direction of plunge (*see* Fig. 83). The gradient of the plunge is calculated as in (a) above, except that the distance between the stratum contours is measured *along the fold axis*.

(c) *The dip and throw of faults:* stratum contours (or structure contours in this case) can be constructed on the outcrop line of a fault plane in the same way as they are on normal bedding planes. The direction of dip of the fault and its gradient are then determined in exactly the same way as (a) above (*see* Fig. 81(a)).

The throw of faults is determined by comparing stratum contour values on either side of a fault. If, for example, (Fig. 81(a)), a particular bedding plane has an altitude of 500 m on one side of a fault, and immediately adjacent on the other side its value is 650 m, clearly the former side has a downthrow of 150 m. The measurement of displacement or shift of a transcurrent fault should be made horizontally along the fault line.

(d) *Vertical thickness of strata:* the vertical thickness of a stratum or a group of strata can be determined by the super-imposition of stratum contours on the map. For example, where a stratum contour constructed on the top of a bed coincides with one drawn on the base of the same bed, then the difference in value between the two represents the vertical thickness of the bed (*see* Fig. 82(a)). It must of course be realised that the vertical thickness of a stratum will vary with its angle of dip. The true thickness, which is measured perpendicular to the bedding, is not affected by this.

(*e*) *Planes of angular unconformity:* discordances in dip between various groups of strata will be very apparent on the construction of stratum contours. In a regularly dipping conformable series the contours usually remain parallel to each other and equally spaced, whereas in examples of angular discordances the stratum contours on the lower group differ from those on the upper group (*see* Fig. 81(*b*)).

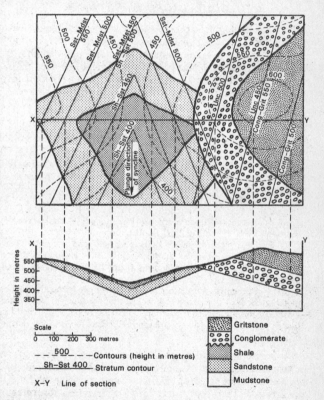

FIG. 83.—*Construction of geological sections from stratum contours.*

(*f*) *The height of a bed at any particular point* can be told from the stratum contours and consequently its depth beneath the surface at any point can also be calculated.

(*g*) *Positions of outcrop of a particular stratum* can be forecast by using stratum contours. Where a stratum contour and a topographic contour of similar altitude intersect there must be a point of outcrop. By plotting all such points and joining them to form a continuous line, a bedding plane outcrop is drawn in.

GEOLOGICAL CROSS-SECTIONS

6. Drawing geological cross-sections. As was the case with map interpretation, the construction of cross-sections can vary from the very simple in elementary problem maps to the extremely complicated in Institute of Geological Sciences maps. For cross-section drawing, geological maps can be divided into two categories, each requiring a slightly different approach:

(*a*) *Maps in which stratum contours are constructed.* Cross-sections can be drawn from the stratum contours which cross the line of section (*see* Fig. 83). The steps involved are as follows:

(*i*) Construct a geographical profile from the topographical contours crossing the line of section. A suitable vertical scale should be selected so that all the detail can be seen clearly without too much exaggeration of either the relief or the dips. It must be emphasised that dips can be made to appear over-steep by too great an exaggeration. In many problem maps it is usual to employ the same vertical scale as horizontal scale.

(*ii*) Using a piece of paper, mark and label the positions where stratum contours cross the line of section. It is often best to deal with one particular bedding plane at a time starting with the uppermost beds.

(*iii*) Transfer the points marked on the paper on to the section plotting them in their correct position and at their correct height.

(*iv*) By joining together the points relating to particular bedding planes, the subsurface strata are gradually drawn in at their correct dip.

(*b*) *Maps with stratal gradients indicated by dip arrows.* Institute of Geological Sciences maps and simplified geological maps often show the dips of the various groups of strata with dip arrows. They may also provide marginal information giving the *true thicknesses* of the beds. Sketch sections across these maps are often not difficult to produce, but very accurate representations of the structure are not always so possible. The steps involved in producing a cross-section are (*see* Fig. 84):

(*i*) Construct a geographical profile in the same way as previously mentioned. For Institute of Geological Sciences one inch to one mile maps an exaggeration of vertical scale is necessary (usually an exaggeration of about ×5 is suitable).

(*ii*) Using a piece of paper mark in the positions where outcrop lines cross the section line. Label these points.

(*iii*) Transfer these points to their correct position on the topographical profile.

(*iv*) Using the values given by the dip arrows near the section line, it may now be possible to draw in the bedding planes from the outcrop points.

(c) *Dip of beds*: *important points to note*.

(*i*) An exaggeration of vertical scale in the profile should affect the dip. The greater the exaggeration, the steeper the dip should appear.

(*ii*) An exaggeration of dip will in turn exaggerate the thickness of beds. When a bed is being drawn in from fixed outcrop points, it follows that the steeper the dip, the thicker it will appear to be.

(*iii*) The *true dip* of beds will only appear on the section if the section line is parallel to the dip direction. Sections drawn obliquely to the dip direction show an *apparent dip* which is always less in value than the true dip. It is possible to calculate the apparent dip from the true dip trigonometrically, but in many cases it can be estimated for the purpose of sketch sections. It must be remembered that on sections at right angles to the dip (strike sections) the strata will appear horizontal. On lines taken between the dip and strike directions, the apparent dip will have a value between that of the true dip and the horizontal. It does not follow, however, that a section taken at an angle half-way between the dip and strike directions shows the strata dipping at half their true dip value. The mathematical relationship is not this simple. To illustrate this a few of the dip values for a section line drawn at 45° to the strike are given below:

True dip	Apparent dip (45° to strike)
20°	14½°
30°	22°
40°	31°
50°	40°
60°	51°
70°	63°

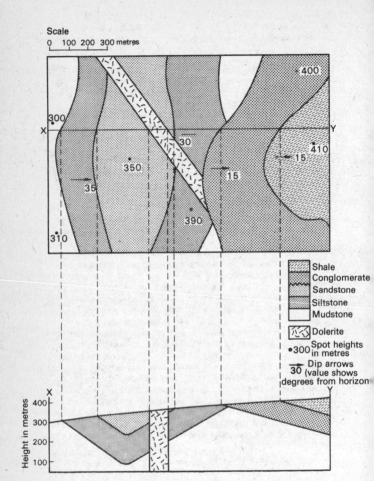

Fig. 84.—*Construction of geological sections from simplified geological maps using dip arrows.*

If the line of a cross-section can be chosen, it should be selected where possible parallel to the dip direction of the strata. At the same time, it is important that it crosses the features of geological importance such as faults, intrusions, unconformities, etc. In fact it may be that two sections are necessary to illustrate the structure of the map.

7. Describing geological maps. If a full description of a geological map is required the following details should be included:

(a) *Title of the area:* Institute of Geological Sciences Sheet No. 238 Aylesbury, scale one inch to one mile.

(b) *Topography:* a short description only is necessary.

(c) *Rock sequence:* the stratigraphical sequence is given on many maps but if it is not then a column of strata should be produced to illustrate this. If possible the thicknesses of the strata should be supplied as well.

(d) *Tectonic structure:* the description of structure is one of the more important elements of map interpretation. It may include:

(i) folding: the types, symmetries, trends, and ages of folding should be discussed;

(ii) faulting: as with folding, the types, throws, trends and ages of faulting should be given;

(iii) joints and cleavage: these are seldom shown on maps but if they should appear, then a description is necessary;

(iv) unconformities: the stratigraphical and geographical positions of unconformities should be noted. Their type should also be identified—oversteps with angular discordance, overlaps, nonsequences, etc. It may be possible in addition to identify the type and degree of erosion or planation involved in the formation of the unconformity.

Small sketch maps showing the positions of the structural elements described are often helpful illustrations in this section.

(e) *Geological section:* an illustrative cross-section should be given. It may be useful to construct this before describing the structure.

(f) *Igneous activity:* a description of types of igneous intrusion and extrusion and their age of formation is required.

(g) *Economic geology:* a short account of the map evidence for economic activities, *e.g.* mining, quarrying, etc.

(h) *Geological history:* this should follow the form of a chronological sequence of events from the deposition of the earliest beds through to the existing erosion of the topography. Only map evidence should be used and stratigraphical general knowledge of palaeo-environments, etc., should not be included. In some cases petrological and palaeontological specimens are supplied with and linked to a problem map. It may be reasonable to deduce some facts about conditions of deposition of strata from these specimens.

PROGRESS TEST 19

1. Give a brief description of the information shown on geological maps. (1)

2. Discuss the ways in which stratal dips may be determined from outcrop patterns. (3–5)

3. What methods may be employed to determine the dip and throw of faults? (3, 5)

4. Describe how the following features may be recognised on a geological map: unconformity, transcurrent fault, dyke. (3, 5)

5. Describe the method of constructing stratum contours. (4)

6. Discuss how cross-sections are selected and constructed to illustrate the structure of an area shown on a geological map. (6)

7. What are the main elements that should be included in the description of the geology of an area? (7)

GLOSSARY

AGGRADATION: The process of building up the land surface by the deposition of material; usually applied to the laying down of alluvium upon a river bed.

ALLUVIAL FAN: The fan- or cone-shaped deposit laid down by a swiftly flowing stream when it enters an open valley or plain and its velocity is suddenly checked; it is particularly characteristic of dry regions.

ANTICLINE: A term applied to stratified rocks bent into an arch-like form; the opposite of a syncline.

ANTICLINORIUM: A compound anticlinal structure; an anticlinal structure in which the limbs themselves are compressed into smaller folds.

APPARENT DIP: *See* dip.

AQUIFER: A porous, water-holding bed of rock, lying between impermeable layers.

ARÊTE: (French) a sharp crested ridge developed between adjacent valley or cirque glaciers.

ASTHENOSPHERE: A zone of the upper mantle (approximately around 100 km from the earth's surface) which is believed to be near melting point. It is regarded as a "weak" layer and one in which motion such as convection is most likely to occur.

AVALANCHE: A mass of snow and ice which slides rapidly down mountain slopes, collecting and carrying large quantities of rock in its destructive descent.

AXIAL PLANE: An imaginary plane bisecting a fold.

AXIS (FOLD): The vertical trace of an axial plane; also used by some as the horizontal trace of an axial plane.

BADLANDS: Elevated, arid regions eroded into innumerable deep gullies separated by upstanding ridges or platforms.

BARCHAN: An isolated, crescentic sand-dune, the horns of the crescent pointing down wind; the dune has a gradual slope on the windward side, a steep concave slope on the leeward.

BASE-LEVEL: The lowest level to which a stream can erode its bed; the final and permanent base-level is sea level.

BEACH: The strip of land, terrace or platform, lying between high and low water marks, formed by sea action; on sheltered coasts it usually consists of sand, on exposed coasts of shingle.

BEDDING PLANE: The surface upon which the material which now forms a rock stratum was originally deposited; such surfaces form planes of weakness.

BERGSCHRUND: (German) the marked cleft or crevice in the snow field of a cirque a few feet away from the rock wall; it marks the zone of separation between stationary snow and moving ice.

BILATERAL SYMMETRY: A symmetry displayed by fossils whereby it is possible to divide the organism by a plane so that one half is a mirror image of the other.

BIOFACIES: The faunal and floral characteristics of a sedimentary formation; often used in contrasting one group of strata with another (*see* facies).

BLOW-HOLES: Holes drilled through the roofs of sea-shore caves as a result of wave action and the compression of air. Water is sometimes spouted through by the force of the rising tide. Same as a chimney.

BOLSON: A basin of interior drainage in arid or semi-arid areas, where infilling, by alluvial fans around its flanks, tends to take place.

BOTRYOIDAL: Mineral form consisting of an aggregation of spheres; resembles a bunch of grapes.

BUTTE: A steep-sided, flat-topped hill standing above the adjacent country because of its greater resistance to erosion.

CAÑONS: (Spanish) deep, narrow, steep or vertical-sided valleys, cut by rivers in horizontally-bedded strata.

CARBONATION: The formation of carbonates by water containing carbon dioxide acting upon rocks containing lime.

CHITIN: A highly resistant horny substance which forms the skeleton of some invertebrate animals; it is a nitrogenous carbohydrate.

CLEAVAGE: The property of splitting along regular and largely parallel planes. *Mineral cleavage* is controlled by atomic structure. *Rock cleavage* results from the flattening of mineral particles and the growth of new flaky minerals during conditions of high stress (*e.g.* during folding). *Slaty cleavage* is the most perfectly developed type of rock cleavage. In this the planes are particularly closely spaced and regular, and they form parallel to the axial plane of the fold.

CORRASION: The vertical or lateral erosion by streams (or glaciers) through the abrasive power of their load of suspended matter.

CORROSION: The solvent action of water.

CORRELATION: The matching of strata (especially in a chronological sense).

CRAG-AND-TAIL: A rock mass which, in obstructing the flow of ice, has been smoothed on the upstream side and has a sloping bank of glacial debris, the tail, on the downstream side.

CRATER: The funnel-shaped hollow at the summit of the cone of a volcano.

CUESTA: A landform produced in areas of gently dipping rocks where a hard, resistant bed caps soft, easily eroded beds; it exhibits a steep scarp-slope on one side and a gentle dip-slope on the other.

CYCLIC SEDIMENTATION: A repetitious sequence of beds of distinctive pattern. An oscillatory cycle has a pattern such as ABCBABCBA. A rhythmic sequence is slightly different and may be represented by

ABCABCABC. One complete unit from such a repetitious sequence is referred to as a *cyclothem*.

DELTA: An alluvial tract, often roughly triangular in shape, formed at the mouth of a river, or in a lake, as a result of the stream depositing more solid material than tidal or other currents can remove.

DESILICIFICATION: The removal of silica or silicates from rocks.

DESICCATION CRACKS: Cracks formed in sediments as a result of drying and shrinkage. A polygonal pattern is produced which indicates the exposure of the sediment to the air.

DODECAHEDRON: A crystal shape belonging to the cubic system; it is formed of twelve rhomb-shaped faces.

DOLINE: (Russian) natural round or elliptical closed hollows found in karst regions (*q.v.*) with a sink-hole into which the surface water disappears.

DETRITUS: Eroded rock debris.

DIAPIR: A dome-like intrusion which migrates upwards by bending and rupturing the rocks above.

DIP: The dip of a bed is the angle between the line of its maximum slope and the horizontal. The *apparent dip* is any observed dip which is not taken along the line of maximum slope and is therefore less in value than the true dip.

DYKE: A sheet of igneous rock which has forced its way more or less vertically through the strata of pre-existing rocks.

EARTH PILLAR: A tall column of earth, which may be 20 or 30 ft (6–9 m) high, capped by a boulder which has prevented the earth beneath it from being eroded away.

EXFOLIATION: The peeling off in layers of rock subjected to alternate expansion and contraction; a feature of mechanical weathering common in desert areas but which can be found under other climatic conditions.

FACIES: A term used to describe the general characteristics of a rock in order to distinguish it from others. It may be used in a variety of ways. Thus "sandy facies" or "alkali granite facies" refer to the lithological character of the rock, whereas "shallow water facies" or "green-schist facies" relate more to conditions of formation. The fossil content of a stratum may also be a diagnostic feature and hence such terms as "graptolitic facies" or "shelly facies" are used. Because of its wide range of usage, facies is now often subdivided into *lithofacies*, *biofacies*, etc.

FAULT: A break, crack or fracture in the earth's crust, along which the strata on one side are displaced relatively to those on the other.

FIORD: (Norwegian) a submerged glacial valley; a long, narrow, deep, usually rectangular-branching, opening along the coast. Characteristically, fiords have an underwater threshold at their seaward end.

FOLIATION: A layered structure often resembling a mass of leaves; foliation results from the parallel growth of minerals and is a common structure in metamorphic minerals and rocks.

GANGUE: The matrix or embedding mass of a metallic ore.

GEANTICLINE: An upward bulging linear area developing within a geosynclinal basin.

GEO: A deep, narrow coastal inlet or cleft which often marks a joint or fault along which erosion has taken place.

GEOSYNCLINE: An elongated trough, resulting from subsidence or the down-warping of the crust, in which a great thickness of sediments accumulate.

GEYSER: A hot spring which intermittently, either regularly or irregularly, spouts water and steam into the air; they are associated with volcanic activity and characteristic of the waning phase of vulcanism.

GONDWANALAND: A former supercontinent composed of the present southern hemisphere landmasses and India. This great continent lay well to the south in the southern hemisphere in pre-Hercynian times but began to split up during the Mesozoic era.

GRABEN: (German) a structure (valley or depression) resulting from the subsidence of a belt of land between two faults, *e.g.* the Rhine Rift Valley.

GREYWACKÉ: a corruption of the German grauwacke (grey rock), a greywacké is a poorly sorted sandstone (muddy sandstone) found in geosynclinal deposits. Greywacké sequences characteristically contain sandstones with graded bedding alternating with shales. They are thought to result from deposition from turbidity currents.

HAFF: (German) a shallow lagoon or bay, enclosed by a long spit of sand, formed at the mouth of a river; characteristic of the southern shores of the Baltic.

HORST: (German) a block or segment of the earth's crust left upstanding as a result of the down-faulting of the land on either side, *e.g.* the Central Massif of France.

HYDRATION: The process of taking up water, as a chemical constituent, by a rock; such water reacts with certain minerals and causes the rock to disintegrate or decompose.

INLIER: An outcrop of older rock surrounded by younger. It may result from upfaulting, upfolding, or deep erosion.

INCISED MEANDER: When, as a result of uplift, the flow of a meandering stream is quickened, the stream recommences downward erosion with the result that the meanders become entrenched between steep, symmetrical sides.

INSELBERG: (German) a residual hill, rising abruptly above the surrounding country, with a round or flat top, the top representing the former plateau level. They occur in semi-arid and arid regions.

KAME: An irregular mound or ridge of fluvioglacial sand and gravel. It may result from deposition at the margin of an ice-sheet or from subglacial deposition.

KARST: A term applied to regions of pervious limestone rocks where surface water disappears underground and the landscape is characteristically bare and gaunt-looking with very little vegetation.

KNICKPOINT: A break in slope, more particularly associated with a river profile.

LACUSTRINE: Pertaining to lakes (*e.g.* lacustrine sediments are those laid down in lakes).

LAURASIA: A former supercontinent composed of the present northern hemisphere landmasses with the exception of India. It came into existence after the Hercynian earth movements but began to split up in the late Mesozoic.

LEACHING: The removal by percolating water of the soluble salts and organic matter in soil.

LEVEE: The natural embankment formed by the deposition of silt when a river floods; it is the highest part of the floodplain. The term is also applied to artificial embankments built to prevent flooding.

LITHOSPHERE: (*i*) A term for the whole of the rock earth-body, including the core. (*ii*) Used in its more restricted sense, it refers to the rigid outermost shell of the earth; this includes the crust and that part of the upper mantle lying above the asthenosphere.

LOESS: (German) a buff-coloured, wind-blown deposit forming a calcareous loam or clay which is extremely fertile; post-glacial in origin, it is homogeneous in character and non-stratified.

MAGMATIC STOPING: A process by which magma migrates upwards. It involves the fracturing and "swallowing" of blocks of roof rock.

MEANDER SCROLL: The land-markings, more particularly distinguishable from the air, of former river meanders now dry and filled by deposition.

MESA: (Spanish) flat-topped hills with steep sides resulting from the dissection of table-lands in arid areas; similar to a butte but larger.

MONADNOCK: An isolated residual hill standing above the surrounding country which owes its origin to the sub-aerial erosion of the pre-existing plain.

MONOCLINE: A tectonic structure in which gently dipping strata are flexed into a vertical or steeply dipping attitude. A monoclinal fold is often associated with normal faulting.

NAPPE: (German) a great recumbent fold which has been torn from its roots and displaced by being pushed forward over a thrust-plane, *i.e.* the plane along which rock masses move horizontally.

NEHRUNG (German) a long sand-bank or spit; they enclose the haffs of the south Baltic coast.

NIVATION: The sapping and weathering of the rock in hollows or valley heads due to the alternate freezing and thawing of snow.

NUNATAK: An isolated mountain peak projecting island-like through an ice-sheet.

OCTAHEDRON: A crystal form of the cubic system consisting of eight equilateral triangular faces.

OOLITE: A rock composed of round grains that display internal concentric layering (ooliths). The texture resembles that of a fish roe. Most oolitic formations are calcareous although some ironstones and other sediments also exhibit the structure.

OOZE: The soft, fine-grained liquid muds which cover the deep-sea beds of the oceans.

OROGENY: A mountain-building movement; a phase involving folding, fracturing and thrusting which results in a change in the level of the earth's crust.

OUTLIER: An outcrop of younger rocks completely surrounded by older. Residual hills resulting from erosion often provide good examples, e.g. Ingleborough, Yorkshire.

OUTWASH PLAIN: The alluvial plain formed in front of a glacier or ice-sheet by the streams issuing from the ice which deposit fine morainic material.

OX-BOW: A lake in an abandoned meander; the remains of a meander which has been short-circuited by the stream cutting across the neck of the loop. Also called cut-offs and mortlakes.

OXIDATION or OXIDISATION: The combination of the oxygen of the air, particularly in the presence of moisture, with certain mineral constitutents in rocks to form oxides.

PANGAEA: The name given by Wegener to the great landmass which formed when the earth's continental masses drifted together in late Palaeozoic and early Mesozoic times.

PEDIMENT: The planed-off rock platforms at the base of scarps in arid mountain areas; the mechanism of their evolution is disputed, but they may well have been developed by the recession of the scarp through weathering and abrasion.

PENEPLAIN: The almost level or gently rolling lowland resulting from long-continued denudation.

PERICLINE: A dome-like or elongated dome-like fold.

PERIGLACIAL: Referring to the fringe area immediately in front of an ice-sheet.

PERMAFROST: The condition of permanently frozen subsoil.

PIEDMONT: Literally, at the foot of the mountains; hence, a piedmont glacier is an extensive ice-sheet at the base of a mountain range formed by the coalescence of several glaciers.

PLANKTONIC: Of floating or drifting habit.

POLJES: (Slav) large flat-bottomed depressions found in karst areas, solution of the limestone playing a large part in their formation.

POTHOLE: A circular hole worn in solid rock as a result of the abrasive action of pebbles being gyrated by swirling water. A term also applied to a large type of sink-hole or swallow-hole found in limestone regions.

PRISM: A term applied to some crystal faces that are parallel to the long axis of the crystal. A prismatic cleavage is one occurring parallel to the prism faces.

PUYS: (French) the conical, volcanic hills of the Auvergne in France.

RAISED BEACH: A bench of beach deposits occurring above the present level of high-water mark; such a narrow coastal shelf is frequently bounded to landward by cliffs. Raised beaches result from either land uplift or the lowering of sea-level.

RECUMBENT FOLD: Folded strata which are so bent over that one limb of the fold structure lies horizontally on top of the other.

REGOLITH: The layer of loose and partially broken rock, sometimes called mantle rock, which covers most of the land surface.

REGRESSION: A retreat of the sea from an area.

RIA: (Spanish) a river valley drowned by the sea as a result of a fall in the level of the land; rias form long, narrow inlets gradually deepening seawards.

RIVER TERRACES: Terrace-like features formed in valleys as a result of river erosion. *Paired* or *rejuvenation terraces* result from a drop in base level due to earth movements or climatic change, while small *unpaired terraces* are often produced by the downstream migration of meanders in soft alluvial sediments. Terraces mark the former levels of the valley floors.

ROCHES MOUTONNÉES: (French) projecting masses of bedrock which have been smoothed and striated by glacial action on the upstream side but which remain rough on the downstream side due to plucking action.

ROCK FLOUR: The finely-ground particles of rock found beneath glaciers and ice-sheets resulting from the abrasion of the bedrock by stones embedded in the base of the ice.

SESSILE: Attached or stationary.

STRAND FLAT: A wave-cut terrace or platform standing slightly above sea-level; relates more particularly to the elevated platform which is especially well-developed along parts of the Norwegian coast.

STRATUM CONTOUR: A horizontal line constructed along a bedding surface. Where the surface is a planar one the contour lines are straight and parallel.

STRIKE: The direction of a horizontal line drawn along any structural or bedding plane. It is expressed as a bearing. In the case of strata the strike is always at right angles to the direction of true (maximum) dip.

STRIKE LINE: Alternative term for stratum contour.

STRUCTURE CONTOUR: A horizontal line constructed along a structural surface (*e.g.* fault, unconformity, ore-body, etc.)

SUB-GLACIAL CHANNEL: Channel eroded beneath ice by meltwater often flowing under strong hydrostatic pressure. The channels are very variable in size and morphology (*e.g.* chutes, tunnel valleys, etc.).

SUBMERGED FOREST: Remains of tree stumps inundated by the sea. Often visible at low tide, they indicate a rise in sea-level relative to the land.

SYNCLINE: The downfolding of strata in the form of a trough.

SYNCLINORIUM: A synclinal structure in which the limbs themselves have been flexed into folds.

TALUS: Scree; the broken, angular fragments of rock found at the base of cliffs or steep slopes, derived from the bedrock by weathering.

TETHYS: A large Mesozoic sea separating Europe and Asia from North Africa, the Middle East and India. This great east-west sea gradually closed and ultimately led to the formation of the Alps and the Himalayas during the Tertiary period by the squeezing and crumpling of the sediments which had accumulated within it.

THALWEG: (German) the longitudinal valley section or profile of a river.

THICKNESS (of beds): The thickness of a stratum measured at right angles to the bedding planes.

TILL: A general term for deposited glacial moraine, but more particularly used to describe the stiff, stony, unstratified boulder clay laid down under an ice-sheet.

TILLITE: A consolidated glacial deposit.

TOMBOLO: (Italian) a tied island; a bar linking an offshore island to the mainland or one island to another island.

TRANSCURRENT FAULT: A wrench fault, i.e. a fault along which displacement occurs in a horizontal fashion. Transcurrent faults can be divided into dextral and sinistral types according to the direction of movement.

TRANSGRESSION (marine): An invasion of a land area by the sea.

TUFF: Fragmental volcanic matter ejected during explosive outbursts; this ash in time becomes cemented to form rock.

TURBIDITY CURRENT: A turbulent aqueous flow containing a thick suspension of sediment. Turbidity currents flow downhill on account of their high density and probably originate from sub-aqueous slumps of unconsolidated sediment.

URSTROMTAL: (German) a term applied to the great, wide, shallow valleys excavated by melt-water during the static phases of the ice-sheet which covered the North European Plain during the Pleistocene period.

WADI: (Arabic) a water-course in arid desert regions which is normally dry but which intermittently carries water after rain storms.

WAULSORTIAN REEF: Calcareous debris bound into knoll-like forms initially, eventually coalescing into more extensive ridges and sheets.

WAVE-CUT PLATFORM: A broad platform or terrace developed as a result of the recession of cliffs through the attack of marine erosion.

WIND GAP: A notch in a ridge, originally cut by a stream, from which the water, formerly flowing through, has now disappeared.

YARDANGS: Parallel ridges found in arid areas which have been undercut by wind abrasion; the ridges often develop fantastic shapes.

ZEUGEN: (German) upstanding tabular masses, resulting from wind erosion in arid areas, which possess hard cappings resting on softer horizontal strata.

APPENDIX II
EXAMINATION TECHNIQUE

1. Four essentials. The examination candidate, if he or she is to be successful, must:

    (*a*) obey the rubric, *i.e.* the instructions;
    (*b*) understand the questions asked;
    (*c*) arrange his or her material satisfactorily;
    (*d*) allocate the time at his disposal equitably.

2. Read the instructions very carefully. It is surprising how often candidates disobey the rubric. The instructions on the examination paper are given to help and guide the candidate, so they should be read most carefully and heeded. Usually 5 or 10 minutes is allowed before the examination begins for the examinee to read the paper. Frequently papers are divided into sections and candidates are asked to answer one or more questions from each section; make sure you answer the correct number of questions required in each section. You will, incidentally, receive no credit for questions answered in excess of the number asked. If a paper has a compulsory question, this must be attempted; it may, indeed, carry more marks than the remaining questions you are asked to answer. If a question asks for maps or diagrams, these should be given since it is possible that a proportion of the marks for the question may be allocated to the illustrations. Try to make your diagrams, sketches, maps, etc. neat, reasonably large, accurate and to the point and do not waste valuable time embellishing them. Do not merely repeat in map or diagrammatic form what you have adequately said in words; diagrams and the like should *add* to your verbal description.

3. Read the questions carefully. First, make sure you understand what the examiner is asking. Study the question carefully: it is often helpful to underline the key or salient points in the question and if you do this it will help you in answering the questions, *e.g.*

"By reference to some *specific regions*, compare the *effects* of ice action in *mountain* areas with those in *lowland* areas."

Secondly, avoid superfluous and irrelevant "padding"; by padding, the student is not fooling the examiner, only himself. Obey the injunction: "Answer the question, the whole question and nothing but the question." Thirdly, allot the time at your disposal equitably so that you do not overwrite on any particular question. Some candidates answer only three questions when four are asked for; they believe that by answering three well they can "make up" for the fourth; this seldom

271

works out in practice. If by chance you do run out of time, jot down in note form the essential points which you intended to bring out in your answer—you will at least gain some credit for this skeleton answer.

4. Presentation. Arrange the material of your answer in an orderly, systematic and logical way. Be precise and to the point; give figures if possible; give appropriate examples. Avoid meaningless phrases and generalised statements. Finally, and this should not need emphasising, write legibly and neatly and use good English, paying some respect to grammar, punctuation and spelling. Avoid using slang, abbreviations, and ungeological expressions. Remember it is the quality, not the quantity, of your answer that matters to the examiner. The good candidate and the one who scores heavily in examinations is the one who can give that little bit extra, whether it be in information, argument or illustrations, which places him above the general run of candidates.

EXAMINATION QUESTIONS

1. Describe the evidence which suggests that there were strong earth movements in Great Britain during the Caledonian orogeny.

2. Describe briefly the geological history of England and Wales during the Mesozoic Era.

3. What sequence of events does an unconformity reveal? Describe an actual example.

4. Describe *either* the Ordovician *or* the Jurassic system with regard to (a) rock types and (b) the fossils.

5. With reference to British stratigraphy, explain the term "cyclic sedimentation."

6. Describe the marine and Old Red Sandstone facies of the Devonian and suggest how these may be correlated.

7. Describe the uses that a geologist may make of either graptolites *or* trilobites. Draw labelled diagrams.

8. Describe and explain the formation of the sediments of the Millstone Grit and Coal Measures in the British Isles.

9. Explain three of the following—metamorphic aureole, unconformity, outlier, evaporites, raised beach.

10. Describe with the aid of labelled diagrams a crinoid—under what conditions and when did they live? Why are they rarely found with graptolites?

11. Compare a regular with an irregular echinoid.

12. With the aid of labelled diagrams, describe either a typical rugose coral or a brachiopod. What deductions can a geologist make from the presence of such fossils in a limestone?

13. Write an essay on *one* of the following topics: (a) the structure of the earth, (b) plate tectonics, (c) faulting.

14. Describe the geological structures found in folded mountain chains and illustrate your answer by means of sketches.

15. Write a brief account of regional metamorphism.

16. Briefly describe and give the origin of the following rocks: (a) conglomerate, (b) greywacké, (c) andesite, (d) granite.

17. Discuss the criteria you would use to differentiate between (a) quartz and calcite, (b) sphalerite and galena.

18. What is understood by the terms batholith and lopolith? What kinds of rock are associated with these intrusions and what rocks and minerals might be produced in the immediate vicinity?

19. Describe the sediments and sedimentary structures which may be

produced by deposition in (a) deltaic environments and (b) marine shelf environments.

20. What is understood by (a) mineral cleavage, (b) hexagonal crystal system, (c) mineral hardness? Illustrate your answer by reference to specific examples.

21. Write an account of *either* stream development *or* glacial deposition.

22. Suggest, with reasons, a classification of volcanic landforms.

23. Describe with the aid of sketches *four* of the following: (a) cirque, (b) river terrace, (c) sandspit, (d) barchan dune, (e) landslide, (f) tor.

24. Discuss the geological factors influencing the form and movement of underground water.

25. Give an account of the geological circumstances leading to the development of an accumulation of oil.

INDEX

A

abandoned meander, 172
abrasion, 192
abrasion platform, 200
absolute dating, 12
abyssal plains, 107
acid igneous rocks, 39–41
adobe, 198
Agnatha, 70
aggradation, 162, 263
aiguilles, 144
Alethopteris, 85
algae, 13, 47, 84–6
allosaurus, 83
alluvial cone, 167
alluvial fan, 167, 263
alumina, 38, 92
aluminium, 17, 19, 38
amianthus, 22
ammonites, 11, 225
ammonoids, 69, 75–6
Amphibia, 70, 82–3
amphibole, 26, 39
Ampthill Clay, 59
Andes, 102, 106, 123, 179
andesite, 40–1
angiosperms, 86
antecedent drainage, 177
anthracite, 44, 48
anticline, 100, 102, 115, 251, 263
anticlinorium, 100, 263
apatite, 24
apparent dip (of beds), 259, 263
applied geology, 5–6, 58, 232–43
aqueous sediments, 51
aquifer, 153, 263
aragonite, 20
archaeopteryx, 84
archaeornis, 84
arenaceous sediments, 44–5
arête, 183, 231, 263
argillaceous sediments, 45–6, 54
arkose, 35, 44–5
asbestos, 22
aseismic zones, 136
asthenosphere, 42, 136, 139, 263

asymmetrical folding, 100–1, 251
Atlantis, 126
attrition, 163, 200
augite, 28–9
Australopithecines, 84
Aves, 70
Azoic era, 11
azurite, 21, 28, 31

B

backwash, 204–5
back-wearing, 159–60
badlands, 62, 263
bahada, 167
barchan, 194, 196–7, 263
bars, 206
barysphere, 93
barytes, 26–7
basal conglomerates, 43
basalt, 40, 42, 117
basic igneous rocks, 39, 42–3
batholith, 37–8, 41, 115–16, 253
bauxite, 235
bedding planes, 48, 60, 156, 247,
    254, 256, 259, 263
belemnites, 76
belemnoids, 69
bergschrund, 264
biofacies, 67, 264–5
biogenic rocks, 43
bioherm, 73, 85
biological weathering, 143, 145–6
biotite, 29
birds, 69
blowholes, 201, 264
blowouts, 193
bolson, 264
bosses, 41, 115, 249, 253
boulder clay, 144, 186, 231
brachiopods, 11, 70, 74–5
brachiosaurus, 83
breccias, 43–4, 57, 222
brontosaurus, 83
brucite, 22
butte, 62, 264

275